CTA at 45
A history of the first 45 years of the Chicago Transit Authority

Public passenger transportation by streetcar, bus or rapid transit train in Chicago traces its origins to 1859. It was begun and developed for 88 years by private entrepreneurs who were successful at first, but were ultimately driven out of business by rising costs, changing urban demographics, the competition of automobiles, and unrealistic political control of fares, routes and service.

To preserve and improve Chicago's transit, rapidly becoming moribund in the postwar middle 1940s, it was decided by referendum that public ownership at arm's length from the political structure would be applied. Thus, in 1947, the Chicago Transit Authority was created as an independent municipal corporation. New approaches were put to work to revive the stumbling, but essential service.

Over the years funding and planning problems as well as growing regional bus and rail transit needs have resulted in expanding the public ownership concept and creating some public funding support at the city, state and federal levels. All the while CTA was being fine tuned to meet the challenges of its changing world. In 1993, as it again faces a major crisis, it is timely to review its colorful history.

Here are some highlights of this process, as recalled by a 50-year veteran of Chicago transit (and past CTA Executive Director).

CTA at 45

by George Krambles and Art Peterson

A history of the first 45 years of the Chicago Transit Authority

Table of Contents

Chapter		Page
	Dedication, Acknowledgment and Foreword	5
1 -	The creation and mission of CTA	7
2 -	Challenges facing the infant	9
3 -	What came before	12
4 -	Policy issues 1945-1970	30
5 -	Policy issues 1971-1991	36
6 -	Streetcars and buses	39
7 -	'L' and subway cars	55
8 -	Signalling on the rapid transit	70
9 -	Maintaining the system	75
10 -	Service control	84
11 -	Battling Mother Nature	90
12 -	Employee training and development	99
13 -	Aurora-Elgin, North Shore and others	106
14 -	Safety and security	114
15 -	Trains in expressways and trains to the planes	118
16 -	The Skokie Swift	125
17 -	Ridership, marketing and fares	129
18 -	Partnership in the metropolitan network	134

Appendix
1947 description of the properties 139

Maps:
	Page
Transit service at CTA inception	6
Principal maintenance facilities	78
Where employees live	102
Where employees report to work	103
Improved rapid transit service for northwest Chicago	119
CTA fares, 1947-1993	130
Local and through transit markets in one corridor	135

Bibliography 141
Index 142

Frontispiece: All-electric PCC rapid transit 6000-class cars, unique in the industry, developed by CTA. Shown during performance tests which confirmed the cost-effectiveness of the design. Ultimately, 770 of these dominated CTA's fleet.

Touhy Avenue, Howard line - 10-20-50

Published for the
George Krambles Transit Scholarship Foundation

CTA at 45
A history of the first 45 years of the Chicago Transit Authority
by George Krambles and Art Peterson

Copyright © 1993 by the George Krambles Transit Scholarship Fund, an Illinois not-for-profit corporation awarding grants to assist men and women in their undergraduate or graduate studies in pursuit of professional careers in the transit industry.
Post Office Box 345, Oak Park, Illinois, 60303-3845, U. S. A.

Library of Congress Catalog Card Number 93-91685
International Standard Book Number 0-9637965-4-2

EDITORIAL TEAM, IN ADDITION TO THE AUTHORS
- Roy G. Benedict, maps, research and indexing
- Norman Carlson, research and project management
- William C. Janssen, research and review
- Jim Walter, electronic color separation and pre-press

The team particularly wishes to thank Paul Kadowaki and Fred G. King, retired CTA managers, for their draft input for Chapters 12 and 14, including statistical material used in the charts, and Kendrick D. G. Bisset, former CTA engineer, for similar help with Chapter 8.

All rights reserved. No part of this book may be reproduced or utilized in any form, except for brief quotations, nor by any electronic or mechanical means, including photocopying or recording, nor by any informational storage retrieval system, without permission in writing from the **George Krambles Transit Scholarship Fund,** for whom this book is published. GKTSF directors in 1993 included Ronald Bartkowicz, Robert L. Benton, Norman Carlson, John A. Darling, Arthur D. Dubin, Arthur L. Lloyd, Arthur H. Peterson, Gene M. Randich and Thomas L. Wolgemuth; Program Administrator, Dr. George M. Smerk.

CTA at 45 is a project of an *ad hoc* editorial and publishing team on behalf of the GKTSF.
Image reproduction and assembly by Jim Walter Color Separations, Inc.
Text preparation through desktop publishing. Maps by Roy G. Benedict Publishers' Services.
Printing and binding by Walsworth Publishing.

Dedicated to Walter J. McCarter
1900-1989
General Manager
Chicago Transit Authority
1947 - 1964
He began a new era for transit in Chicago

ACKNOWLEDGMENT

Obviously, this study could not have been undertaken without the full, heartfelt cooperation of the people of Chicago Transit Authority. This was initially made available through Alf Savage, who suggested this book in the first place. Many other CTA people, including active employees, retirees and friends, contributed significantly to the result. We especially want to thank John Santoro and Celestine Offett, respectively Secretary and Assistant Secretary to the Chicago Transit Board, Harold R. Hirsch, Senior Manager of Operations Planning, Supervisor Lillian Culbertson (recently retired) and her associates in CTA's Anthon Memorial Library, and Roy A. Kehl, a dedicated student of electric railway history who, as a volunteer, assists the library in organizing its impressive photographic and historical archives.

FOREWORD

The selection of the subject matter of this book was spurred by the 100th Anniversary of rapid transit in Chicago, which began with steam-powered service on the South Side 'L' on June 6, 1892, and by a perceptive suggestion in the fall of 1990 from then-Executive Director Alfred H. Savage of CTA that we tackle a publication to tell the Chicago Transit Authority story, as contrasted to a general history of transit in Chicago. Both reasons are more than adequate justification for revisiting the topic which a Bulletin of the Central Electric Railfans' Association first treated more than a half-century ago on the occasion of an inspection trip over the Chicago Rapid Transit Company's Garfield Park-Maywood-Westchester line on Lincoln's birthday in 1939.

But the early history of transit in Chicago is not the purpose of this volume, instead, as already indicated, only enough historical background is presented to bring the reader into the setting for the creation of CTA in 1945 and its actual implementation starting in 1947. As CTA completes its 45th year, 1992 becomes the closing year of this book, marking as it does a turning point during which the CTA begins a comprehensive reorganization. Under the leadership of President Robert P. Belcaster (replacing Alf Savage) CTA is taking stock of itself and is seeking to adapt to changing demands for service and shrinking of public funding resources. The changes to be brought about under the new regime are only just beginning as this is written, but they promise to be significant in shaping CTA to enter the 21st century. A tough part of this assignment will be to avoid downsizing the city itself while downsizing its transit services.

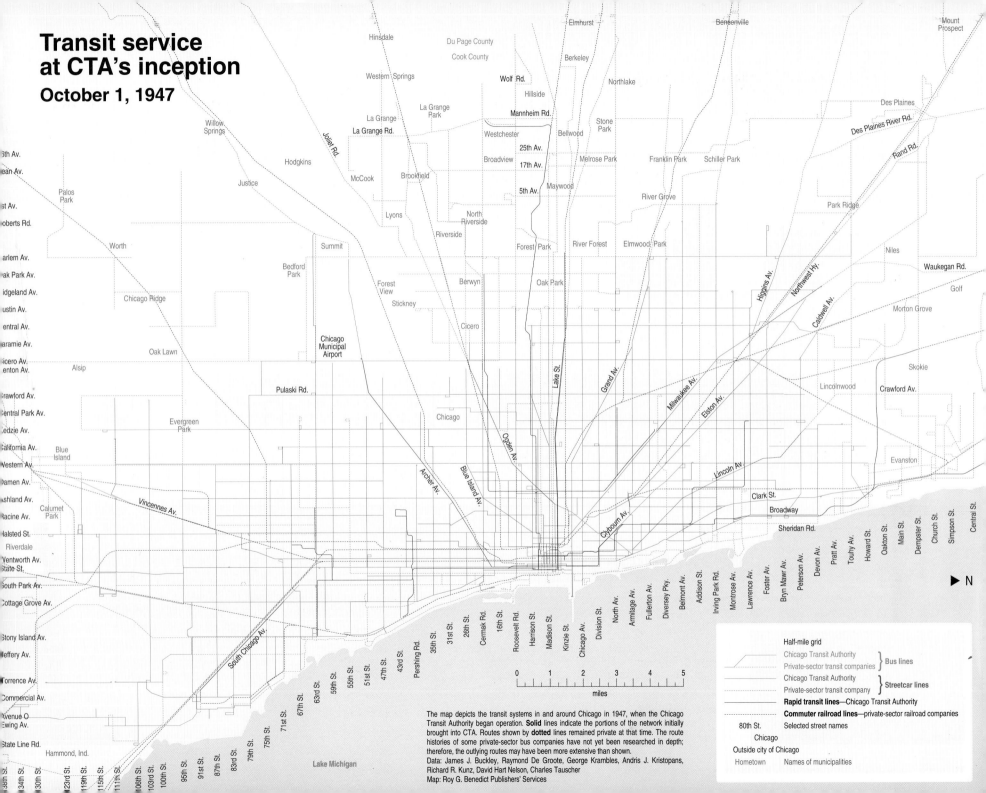

Chapter 1 - The creation and mission of CTA

Chicago Transit Authority was born in 1945, a year in which public transportation in Chicago was provided mainly by streetcars and rapid transit trains.

The street railway system, which included a few trolley bus and motor bus routes, was operated under the name *Chicago Surface Lines* but actually was comprised of five different entities: Chicago Railways Company, Chicago City Railway Company, Calumet and South Chicago Railway Company, Chicago and Western Railway Company, and the Southern Street Railway Company. Due to persistent financial distress, all of these were being managed by trustees or receivers appointed by the District Court of the United States.

The rapid transit lines, known as the *Chicago Rapid Transit Company* and the Union Consolidated Elevated Railway Company, were also being operated by trustees appointed by the Court. Five of CRT's routes extended beyond Chicago through contiguous suburbs and into farther layers of adjoining villages, unlike the CSL, whose routes generally did not go beyond Chicago city limits.

The Chicago Motor Coach Company owned and operated a separate, competing system of motor bus lines in parts of Chicago. The metropolitan area out to 50 miles from downtown was also being served by thirteen companies operating commuter trains, one operating streetcars, and a number running buses. In 1945 these were still in varying degrees of solvency and their inclusion was deferred from the startup of public transit ownership.

The CSL had operated by authority of ordinances passed by the City in the period 1907–1909, all of which had expired by July 15, 1938. The last of similar ordinances granted to CRT were expiring in 1945.

The physical plant and equipment of both CSL and CRT were rapidly approaching obsolescence and urgently needed renewal. What old age hadn't claimed, wartime overloading had! The trustees were barely able to pay current wages and bills, let alone fill needs for modernization or for expanding metropolitan transit service through system growth.

Five major attempts to reorganize the streetcar and rapid transit properties under private ownership had been unsuccessful. After comprehensive debate in the public arena, unification under a single municipal corporation emerged as the only logical alternative.

Therefore, by Act of the General Assembly of the State of Illinois approved April 12, 1945, the *Chicago Transit Authority* was created as a political subdivision, body politic and municipal corporation, to acquire, own and operate a transportation system in the metropolitan area of Cook County.

The Act prescribes, among other things, the objectives, the governing structure, the administrative responsibilities and the permissible financing methods. It granted no taxing powers, but it did exempt CTA from paying certain fuel and sales taxes. It also conferred on CTA the right of eminent domain.

A companion Ordinance of the City Council of Chicago grants to CTA, among other things, the exclusive right to own and operate a comprehensive unified local transportation system within the city. The Ordinance spells out the specific facilities to be included and the methods of compensation for those provided by the City. It identifies certain needs for modernization. It itemizes the segments of streets on which transit service is authorized, including those where future needs were anticipated as well as those then used by CMC, whose acquisition by CTA occurred in 1952.

The Act and the Ordinance were both adopted by a six-to-one majority of the electors of Chicago in a referendum on June 4, 1945, when suburban Elmwood Park also approved.

The governing and administrative element within CTA is the *Chicago Transit Board.* The Board consists of seven members, three appointed by the Governor of the State of Illinois with the advice and consent of the Senate, and four by the Mayor of the City of Chicago, with the advice and consent of the City Council. The Governor's and Mayor's appointments are each subject to the other's approval. Board members have seven-year terms, staggered to reduce the likelihood of abrupt change in policy. The Board elects a Chairman from its membership. The first Board members took their oaths of office on September 1, 1945.

Principal responsibilities of the Board include determining which routes shall be operated by bus, rail or other mode, the levels of service and the fares. They must also approve purchases, sales, rentals, leases and contracts.

To direct the day-to-day operations of the system, the Board is charged to select an individual of recognized capability as a transit manager to become the *Executive Director*[1] (originally termed *General Manager*). It also chooses a *General Attorney.*

The State and Municipal legislation provided no initial grant, tax allocation or taxing power for the acquisition of assets, nor for their modernization or extension. For these purposes, CTA instead was authorized to issue revenue bonds secured by its own earnings. To do so, it

---

[1] After March 1992, *President.*

first had to negotiate a Trust Agreement for the sale of $105 million of bonds and then buyers had to be found.² After these chores were completed, CTA was able to purchase CSL and CRT on October 1, 1947 and to take over running of the buses, cars and trains.

Jumping ahead to complete the story of CTA's creation, in 1952 a further $23 million in revenue bonds was sold, $16.4 million being used to purchase the *Chicago Motor Coach Company,* while the remainder was used for capital improvements. In 1953, a final issue of $7 million was used to purchase a portion of the Chicago Milwaukee St. Paul & Pacific Railroad over which the North-South and Evanston routes were operated. For $135 million, all raised from revenue bonds sold to the public, CTA became Chicago's transit agency. There were no grants in those days and no tax money went into this monumental acquisition.

◇ ◇ ◇

2) For the description of CTA and its initial business plan prepared by the consulting engineer for inclusion in the prospectus offering the $105 million Revenue Bonds Series of 1947 please refer to the Appendix.

105,000 certificates like this comprised the Revenue Bond issue Series of 1947 with which the Chicago Surface Lines and Chicago Rapid Transit Company were acquired. Note the signatures of Chairman Ralph Budd, I. L. Porter, Treasurer and W. W. McKenna, Secretary. These bonds, also those issued in 1952 and 1953, were redeemed in the middle 1970s with the transition to public funding assistance through the Regional Transportation Authority.

Krambles archive

Chapter 2 - Challenges facing the infant

Privatization, one of today's buzzwords, is sometimes touted as a panacea for the "ills" of subsidizing transit. On the other hand, private ownership is sometimes castigated for the problems which brought public transportation to the verge of collapse by the 1940s. Scholarly examination of the record shows that investors in the streetcar and 'L'[1] systems of Chicago ended up unintentionally subsidizing car riders' fares to the tune of nearly $300 million in the 40-year period from 1907 to the 1947 takeover by CTA.

The total bonded debt, with accrued interest, on the Chicago Surface Lines companies as of October 1, 1947 exceeded $201 million. Besides, there were stocks outstanding accruing to an estimated $11 million.

The position of the 'L' was even more distressing, with unpaid interest of more than $64 million and unpaid principal exceeding $57 million. Outstanding debts, interest, stocks and accumulated dividends on preferred stock, plus an accrued loss on a 1938 reduction in common stock, came to a total exceeding $168 million, against which the trustees in bankruptcy had only $19 million with which to pay. Net loss to investors: over $149 million!

All of these numbers should be considered in terms of dollars of the 1940s, perhaps much more than ten times their relative value today.

Reasons for the dismal situation include:[2]

√ - *poor public relations going all the way back to the 1890s, making transit a veritable political football*
√ - *failure to convert low traffic routes to more economical technologies such as one-person buses*
√ - *failure to resolve duplication of routes between competing carriers*
√ - *inability to merge duplicative management and administrative structures*
√ - *inflation, and*
√ - *reluctance of regulatory agencies to permit the operating companies promptly:*
  - *to set fares adequate to generate the funds needed for updating methods, improving equipment and reimbursing investors, and*
  - *to modify routes, close weak stations or decrease service in response to changing land use or new transportation alternatives, and*
  - *to discontinue lightly used night and weekend services.*

An astronomical post-war increase in automobile ownership and the flight of families to new homes in the suburbs were also factors that severely reduced transit market share. The advent, after the war, of the five-day week eliminated home-to-and-from-work trips on the sixth day. This by itself would amount to a loss of more than 14% of transit's ridership. Even the introduction of home television comes in for some of the blame for the shrinking of transit's market share, causing people to stay home when formerly they had gone out several nights a week to find entertainment.

These factors were readily apparent to those who had put together CTA as well as to the Board and General Manager. To begin the process of offsetting those negatives, an intensive short range planning effort was undertaken "at forced march" to utilize the self-regulating powers of the new organization. "Downsizing" was not a word in common vocabulary, but a strong dose of it was in order, but simultaneous to the pruning of dead routes there were needs for new and extended ones.

CTA's planners got busy. Data were collected and plans were developed for the elimination of duplicating services and weak 'L' stations, for the general speeding up of 'L' service and to expedite the replacement of streetcars by buses. Equipment engineers evaluated designs and prepared specifications, while purchasing staff solicited best prices for the hundreds of buses that would soon be needed to implement the service plans.

Within a few weeks the agenda of frequent Board meetings was filled with recommendations and the Board authorized most with "all deliberate haste." In 1945 the public had voted for action but it had taken two years to sell the revenue bonds so that CTA could get going. It was now indeed to be a busy time for CTA!

◊ ◊ ◊

---

[1] '*L*', a contraction of *elevated*, is the popular term in Chicago for *rapid transit,* applying equally to its elevated, surface and subway portions.

[2] These problems are endemic to transit and recur even today.

Members of the Chicago Transit Board since its inception

Chairman	Appointed by	Term served	Additional board members	Appointed by	Term served
Philip Harrington	Mayor	1945 - 1949	William W. McKenna	Mayor	1945 - 1971
			John Quincy Adams	Governor	1945 - 1947
			George F. Getz, Jr.	Governor	1945 - 1947
			Philip W. Collins	Governor	1945 - 1958
			Irving L. Porter	Mayor	1945 - 1950
			James R. Quinn	Mayor	1945 - 1976
			Frank McNair	Governor	1947 - 1949
			Guy A. Richardson	Governor	1947 - 1954
Ralph Budd	Mayor	1949 - 1952	John S. Miller	Governor	1949 - 1956
			John Holmes	Mayor	1952 - 1958
Virgil E. Gunlock	Mayor	1952 - 1963	Werner W. Schroeder	Mayor	1954 - 1960
			Bernice T. Van Der Vries	Governor	1957 - 1971
			Edward F. Moore	Governor	1958 - 1963
			Joseph D. Murphy	Mayor	1958 - 1969
			Raymond D. Peacock	Governor	1961 - 1970
			James E. Rutherford	Governor	1963 - 1969
George L. DeMent	Mayor	1963 - 1971	Ernie Banks	Governor	1969 - 1981
			Wallace D. Johnson	Governor	1970 - 1976
			Clair M. Roddewig①	Mayor	1970 - 1975
Michael J. Cafferty	Mayor	1971 - 1973	Donald Walsh	Mayor	1971 - 1978
			Lawrence C. Sucsy	Governor	1971 - 1979
Milton Pikarsky	Mayor	1973 - 1975	Edward F. Brabec	Mayor	1975 - 1979
James J. McDonough②	Mayor	1976 - 1979	Mathilda A. Jakubowski	Governor	1976 - 1979
			James P. Gallagher	Mayor	1978 - 1986
			John J. Hoellen	Governor	1979 - 1990
			Howard C. Medley, Sr.	Mayor	1979 - 1989
			Nick Ruggiero	Governor	1979 - 1987
Eugene M. Barnes	Mayor	1979 - 1982	Michael I. Brady	Mayor	1981 - 1985
			Jordan J. Hillman	Governor	1981 - 1987
Michael A. Cardilli	Mayor	1982 - 1986	James I. Charlton	Mayor	1986 - date
Walter Clark	Mayor	1986 - 1988	Natalia Delgado	Mayor	1986 - 1992
			J. Douglas Donenfeld	Governor	1987 - date
			Milton Holzman	Governor	1987 - date
Clark Burrus	Mayor	1988 - date	Arthur F. Hill, Jr.	Mayor	1989 - date
			Kim B. Fox	Governor	1990 - date
			Guadalupe Reyes	Mayor	1992 - date

① - Roddewig was Acting Chairman during period following death of DeMent. Other Vice-Chairmen served similarly.
② - McDonough was Acting Chairman from Feb. 1976, Chairman from Oct. 1976 - 1979, and Board member through 1980.

Chief executive officers

Walter J. McCarter	1947 - 1964	General Manager
Thomas B. O'Connor	1964 - 1973	General Manager
M. Pikarsky, J. Aurand, T. Hill, P. Kole & G. Krambles	1973 - 1975	Management committee
George Krambles	1976 - 1980	Genl. Mgr., Feb. 1976; Executive Dir., Oct. 1976
Theodore G. Schuster	1981 - 1982	Executive Director
Bernard J. Ford	1982 - 1985	Executive Director
Robert E. Paaswell	1986 - 1989	Executive Director
Bernard J. Ford	1989 - 1990	Engagement Manager
Alfred H. Savage	1990 - 1992	Executive Director
Robert P. Belcaster	1992 - date	President

6-21-45 Krambles archive

The CTA statute specifies that the governing and administrative body, Chicago Transit Board, shall consist of seven members, "residents of the metropolitan area and persons of recognized business ability". It specifies requirements for their appointment and confirmation as well as terms of office and duties. The first board, shown above, included, left to right, seated: Philip Harrington (Chairman), James R. Quinn, William W. McKenna, standing: Irving L. Porter, Philip W. Collins and John Q. Adams. Member George F. Getz was not present for the photo. Their first tasks were to sell the revenue bonds which would buy the CSL and CRT properties and to recruit a General Manager to bring together into one team the then-competing groups.

Chicago's Mayors, who nominate four of the seven-person board and thus select the chairman, have always shown keen interest in transit and participate personally, encouraging improvement, extension and public support. Left to right, about to enjoy a test run of prototype high-performance cars, are CTA Chairman Virgil Gunlock, Mayor Richard J. Daley, Alderman J. E. Egan, CTA board members W. W. McKenna, P. W. Collins and J. R. Quinn, Alderman O. F. Janousek and CTA board member J. Holmes.

North Water Stub - 10-3-55 - Krambles archive

Posing for a formal portrait with the Board with whom he served for most of his final year with CTA is Executive Director George Krambles. Left to right seated: James J. McDonough, Chairman Eugene M. Barnes, and Mathilda A. Jakubowski. Standing: Krambles, Howard C. Medley, Sr., James P. Gallagher, Ernie Banks and Nick Ruggiero.

CTA Board Room - 8-1-79 - CTA

Chapter 3 - What came before

The CTA that began its corporate life on September 1, 1945 and its operating life on October 1 two years later certainly took over a mammoth transit system, the second largest in North America. In round numbers there were 5,000 vehicles running over 2,000 miles of route blanketing the city with service within three blocks of virtually all of the population. Two-thirds of the lines operated around the clock every day. More than three million rides were taken on an average weekday.

It was also a mature system. For example, where the reasonable life expectancy of a rail vehicle was 25 years, the CSL streetcar fleet had an average age of 28. CRT's fleet averaged 40 years and two-thirds of it was of wooden construction, obsolete already for 30 years! Shops, yards, stations and power equipment were of comparable vintage. Wartime overload traffic demands and shortages of new materials were still taking an accelerated toll of wear-and-tear from all of the physical plant.

The historical background before CTA is a fascinating story whose documentation has inspired many writers and photographers over the years. In *A History of the Yerkes System of Street Railways in the City of Chicago*, 1897, one reads:

"A city is known by the character of its street railways. They are a never-failing index of its enterprise. Intramural transportation will keep pace with the liveliest community; in nine cases out of ten it will lead the way. So it has been in the history of Chicago street railway construction of the past decade; indeed, to the pioneer work of railway capitalists in all stages of its history is due very much of the unparalleled development of this unequalled city. Take from Chicago the rapid transit facilities, and she would be crippled beyond repair. Relegate her to the methods and motive power of a dozen years ago, and half her business interests would be paralyzed . . ."

The Yerkes history goes on to report that in 1858 the main streets were unpaved and at times almost impassable with mud. Beyond the south city limit at 22nd Street stretched a vast bare prairie, with only here and there a dwelling. Two horse **omnibus** (later 'bus, then bus) lines carried people from the railroad stations, but since 1855 the subject of street railways had been agitated. In 1856 the Council awarded a franchise to Roswell B. Mason and Charles B. Phillips to build a line on State Street from Randolph to 22nd (today's Cermak) and one on Dearborn and Franklin to Fullerton, the north city limit at the time.

Unfortunately, there was a financial panic the next year and the project died, but in 1858 the Council made another grant, this time for lines on State, Cottage Grove, Archer and Madison. The main promoter of this scheme was Frank Parmelee, who five years earlier, using five or six old yellow buses belonging to hotels, had started a service transferring passengers between railroad stations. Next, a line was put on State Street between Randolph and 12th (now Roosevelt) for transportation of the general public and before long there were 70 coaches, 250 horses and 200 men involved.

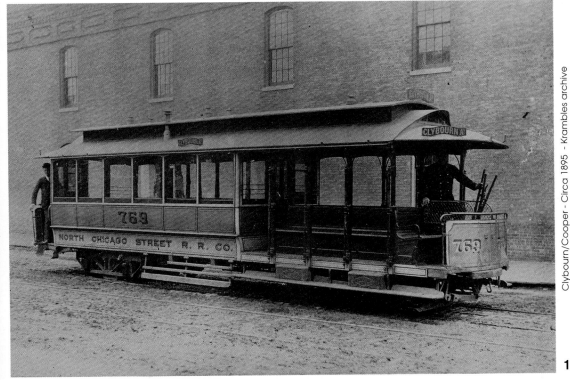

Clybourn/Cooper - Circa 1895 - Krambles archive

Chicago had some of the largest cable cars anywhere, for example, North Chicago Street Railroad #769. It was assigned to the 3-mile long Clybourn route, opened in 1891, utilizing the troublesome (and ironically named) *Low & Grim* **cable top grip. The line was converted to trolley cars October 21, 1906.**

Even though State was a main street, it was very muddy. Along its center was a corduroy road of loose planks, either side of which was bottomless black mud. Quoting the Yerkes history, *"If a coach chanced to run off the planking, it took two days, with ropes, men and teams to haul it up again. Even on the planking the travelling was none too good. J. F. Johnson, of the Chicago City Railway Company, says he remembers well the squirts and streams of black mud that shot up between the boards as the 'buses went rattling over them, the oozy fountains often playing many feet higher than the tops of the coaches themselves, completely covering them with liquid blackness."*

Chicago Surface Lines

Running on rails would lift the omnibuses out of this mess, and Parmelee proposed a horse railway, but by then the Council had heard so many unfulfilled claims for street railways that it remained skeptical. To convince them of his good faith, Parmelee ordered from Eaton, Gilbert & Company of Troy, New York, four horse cars to be sent at once.

"They were made in a hurry and shipped by the Lake Shore & Michigan Southern Railway to Chicago, being unloaded at Twelfth Street. Then the famous first trip of a street car in Chicago was made. The four cars were unloaded one by one and hauled by teams over the plank road in State Street to Randolph, where they were left standing as an object lesson to convince the council of Mr. Parmelee's good faith. The franchise was at once forthcoming, and on November 1, 1858, at State and Randolph Streets, ground was broken for the first street car line in Chicago."

The line, single track at first, opened for business in early 1859 and reached 12th Street by April 25th. On February 14, 1859, by a special Act the legislature of the State of Illinois approved the franchise granted by the City Council to Parmelee's group to operate streetcars in the south division of the city for 25 years. Of course, the omnibus operators strongly opposed the street railways and a fare war was fought pushing prices down until tickets were almost given away to get patronage. Parmelee himself sold out by 1864.

In the end, the streetcar companies were successful, the 'bus lines being sold out or withdrawn. Development on the north side of the city had also started in 1859 under the North Chicago Street Railway Company and on the west side by other companies such as the Chicago West Division Railway. The great Chicago fire of 1871 inflicted heavy damage on the streetcar system, but repair started *"no sooner had the embers ceased smoldering."*

Fares again climbed and were being sold at 6¢ immediately after the Civil War when an internal revenue tax had been established. They dropped back to 5¢ when the tax was lifted and stayed there until August 1919.

There was little change in the streetcar system from 1865 to 1875, but then expansion resumed apace. By 1880 it had become so extensive that the horse cars were entirely inadequate. Horses moved slowly, in winter the cars were cheerless and cold; hay thrown on the floors did little to retain what heat radiated from riders. The animals were a public nuisance and health hazard; their food and care were costly. They couldn't be worked pulling a streetcar for more than a couple of hours at a time. They were prone to illness.

In the search for an alternative, C. B. Holmes, president of the Chicago City Railway, was greatly impressed by the technology that seemed to be proving successful in San Francisco, where cars were towed by gripping a continuous flexible cable running on pulleys buried in a conduit under the street. The cable moved constantly, powered by coal-fired steam engines in power plants along the route.

Holmes' company aggressively pursued the concept after obtaining exclusive rights to use the San Francisco patents in Chicago. It got our city's first cable line, on State between Lake and

10-9-1893 - W. C. Janssen collection

Rarely in the life of a transit service is there opportunity for such record traffic as was experienced by the Cottage Grove cable line of Chicago City Railway on Chicago Day at the 1893 World's Columbian Exposition. Capacity crowds were also served that day on steam suburban trains of the Illinois Central Railroad as well as on the new South Side Rapid Transit line.

Circa 1907 - by F. E. Borchert, J. L. Diaz collection

Annexation of four Townships into Chicago in 1889 placed considerable strain on postal pick up/delivery. Street conditions and distance thwarted the progress of horse-drawn vehicles, so the city's street railway system was used to expedite mail carriage. Originally using cable car trailers as its mail cars, the post office switched to self-contained electric cars in 1899. At one time, six routes covered major sectors of the city, but by 1915 streetcar mail service ended. The rolling stock became work cars.

Initially, each electric railway generated its own electricity. Shown is the boiler room of Hobbie Street Power House. Between 1896 and 1904, from its location along the north branch of the Chicago River, this station fed direct current to the the north side streetcar area. Coal flowed in from the hoppers above and ash was withdrawn at the bottom, with many labor-intensive tasks in the process. Ultimately, the railways' power demand became the base of the utility business and small company-owned power houses were replaced by central station power supplied by Commonwealth Edison. Alternating current now is distributed from CE's network to transit substations scattered over the system, where it is converted to direct current.

North Branch Chicago River/1030 north - 1900 - Krambles archive

The predecessors of CTA suffered from competition in areas of the city with heavy potential ridership, but also from overextension into some of the underdeveloped sectors, a diseconomy compounded by uneconomic crewing requirements. For example, the Ewing-Brandon line, to reach a steel mill and the outlying community of Hegewisch, traversed the Wolf Lake wetlands using two-man cars (or flagmen) to protect steam railroad crossings.

9-15-46 - J. J. Buckley

Circa 1910 - Krambles archive

Summer cars were only briefly popular in Chicago. Calumet & South Chicago #603, formerly South Chicago City Railway trail car 204, was unusual for its single-end, center aisle layout, rather than the conventional plan with five-across benches and continuous running boards along the sides. C&SC had been formed in 1908 by the merging of SCCR with the Calumet Electric Street Railway. CESR had been Chicago's pioneer in electric traction, introduced on October 2, 1890, just five months after that company's incorporation.

Elston/Addison - 4-9-23 - Krambles archive

Van Buren/Franklin - 6-10-47 - T. H. Desnoyers

Nine garages serve Chicago's buses in 1993, but, when CTA took over, CSL had sixteen operating locations, most dating back to 1908. Chicago Electric Transit Company opened its Elston carhouse on December 26, 1894 and it served until 1951. The barn could hold 95 double-truck cars, 49 of them inside. When this wasn't enough, cars were stored on Elston Avenue, and revenue trips had to single-track around them.

In the early 20th century when navigation was predominantly commercial, not pleasure traffic and far heavier than today, the Chicago River with its north and south branches effectively barred access by cable car to the central business area from the north and west sides of the city, Yerkes overcame the problem by taking over the under-river tunnels on LaSalle and Washington and building the Van Buren tunnel new. Although regular routing through the latter ended in 1924, it was often used as a reroute and its heavy grades were useful in training new motormen.

21st Streets, built and in service by January 28, 1882. By 1888 the Chicago City Railway was using 300 cable trains in peak hours. On October 9, 1893, Chicago Day at the World's Columbian Exposition, 700,000 riders were carried, mainly on cable trains.

The North Chicago Street Railroad had been acquired by Widener-Elkins interests, a Philadelphia investment syndicate, in collaboration with Charles T. Yerkes who had come to Chicago from Philadelphia in 1881 at the age of 44. Although he left Chicago before 1900, that was long enough for him to become a controversial character—a gifted entrepreneur who "gave the city transportation when it needed it, awoke the enmity of powerful men of finance, bludgeoned his way to success through backroom lobbies and threats, and all in all led a most colorful life", memorialized in Theodore Dreiser's novel, *The Financier.*

Yerkes decided to employ cable technology to upgrade the horsecar lines, which crossed the Chicago River on movable swing bridges. Navigation was constantly delaying the company's cars 15 minutes or more at a time. Movable bridges were unadaptable for the mechanics of cable traction. To cross them, grips would have to be removed while horses towed the cable cars over the river and then the grips would have to be reinstalled. Yerkes turned this problem to advantage over all other street traffic when he conceived the idea of rebuilding an old pedestrian tunnel under the river at LaSalle Street as a crossing for cable cars. South of the tunnel his cars were then fed into a loop of several streets

Competing services before CTA were illustrated by CSL Armitage-Downtown car 601 rounding the vertical curve from Clinton street to enter the Washington tunnel, the trunk line link for streetcars from the northwest side. At the same time, CMC double decker 665 in the paralleling street traffic lane prepares to cross the river via the Washington bridge. The tunnel was fitted with color light signals; they did not indicate track occupancy but were timing devices used to indicate safe downhill running speed to motormen.

providing distribution through downtown, although Holmes' City Railway blocked access to the more desirable east part of the area.

To avoid having to license San Francisco's proprietary design, the North Chicago company opted for a grip design that had been developed for Philadelphia, albeit not very successfully according to some reports. Characteristically claiming the superiority of the Philadelphia scheme, Yerkes rapidly completed construction and got into operation on March 26, 1888. At one time, a fleet of 157 grip cars plus 515 trailers was inventoried.

Yerkes next acquired control over the west side lines. In the words of the Yerkes history, *"It needed but the touch of this master hand to transform the entire West Division from a desert of horse railway tracks into an oasis of rapid transit by cable traction."* Car lines from the west crossed the south branch of the Chicago River. Copying the successful LaSalle Street strategy, Yerkes acquired the little used Washington Street tunnel and converted it for cable cars and then built a new tunnel near Van Buren Street (with grades up to 10%) as access for a Blue Island Avenue line.[1]

The troublesome cable grip of the North Chicago company was not used for these lines. Instead, one new design was used for the Madison Street and Milwaukee Avenue lines and another for the Blue Island Avenue route, so the Yerkes system wound up with three non-compatible grip schemes.

All told, about 30 miles of double track and 200 grip cars were included in the west side cable system, which began its first service on Milwaukee Avenue on June 7, 1890 and on Madison Street on July 16.

Chicago's venture into cable cars was the most extensive in America, far exceeding that of San Francisco. It also represented the peak of complexity for any mechanical transportation concept in history.

But by 1888, before the last of Chicago's cables was complete, a new technology, the electric trolley car, was put in revenue service in Richmond, Virginia. In the next year or two it was in use throughout the world, foredooming the era of the cable car.

However, there was initial reluctance to try electric traction in Chicago, as picturesquely documented in the Yerkes history: *"So bitter was the opposition of the people of Chicago to electric roads that the vastly improved service which electricity was destined to give the city was long postponed for no other cause than blind prejudice. Not until nearly all the suburban towns had enjoyed the benefits of the rapid running trolley cars for many months was this system reluctantly tested inside the Chicago city limits, and even then the change from horses on outlying connecting lines to electric traction was made with great hesitation on the part of abutting property owners. Even [in 1897] on a prominent North side residence street, the old lumbering, out-of-date horse car system is still in vogue, simply because the property owners along that avenue have steadfastly refused to allow their precincts to be invaded by trolley."*

Desperate to replace animal traction but with something more practical than cable, a number of ideas, including steam, soda, compressed air and gas motors, but it was evident that electricity was the answer.

---

[1] The original LaSalle tunnel was built 1869-1871; the Washington tunnel in 1867, and the Van Buren tunnel in 1890-1894. Later, after the river flow had been reversed, its level lowered. In 1904 Congress declared the tunnels to be obstructions to navigation. All were rebuilt by 1912. They still exist, but are now abandoned.

Soldier Field Gateway - 1933 - Krambles archive

In an effort to avoid complaints about the poles and exposed overhead wires needed for *trolley* cars, an underground conduit system was tested on a loop of track in the vicinity of the McCormick Theological Seminary (near the intersection of Fullerton and Halsted streets). Concluding that it was too costly and vulnerable to failure and vandalism Yerkes proclaimed, *"One ride on the trolley cars was sufficient to convince the people that nothing better could be had."* No conduits for him!

By the early 1890s, ordinances enabling electrification finally did pass the City Council, and within a year or two many miles of electric railway were operating in Chicago.

The Century of Progress Exposition of 1933-1934 spread along the lake front from Roosevelt Road to 37th Street. Heavy crowds were served by streetcars, trains and buses. Above is the north gate terminal of CSL. In the background is the viaduct (over Illinois Central Railroad's steam and electric tracks) linking the terminal to the Roosevelt car line, completed by the city two months after the fair opened. In the meantime CSL riders walked in from the 31st car line or rode in on the Cermak line to gates at either 23rd or 18th. Below is a homeward bound crowd with Soldier Field in the background and CMC buses on 14th Boulevard.

Milwaukee/Hubbard - 1-22-40 - Dept. of Subways & Superhighways

Construction of subway directly beneath important Surface Lines Milwaukee Avenue routes necessitated temporary rerouting of car tracks and disruption of street traffic, even marooning some buildings. At 11th/State Streets temporary tracks took streetcars along the alley right-of-way under the South Side main line 'L'.

Among the largest of the old barns inherited by the system is Archer, with a capacity, as rated in 1922, of 372 double-truck streetcars, all but 43 inside, under cover. It is one of four carbarns that were built to specifications set by the Board of Supervising Engineers, a non-political office established, with renowned transit engineer Bion J. Arnold as its chairman, to administer the 1907 traction ordinances.
It is shown above as it appeared in 1941 and below as in 1948 after minimal conversion to serve buses. Although further remodeled since with wider doors and other improvements, it is now overdue for total replacement to include modern concepts of garage design. These would include complete, secure enclosure of revenue collection, vehicle servicing, circulation and storage activities, with enhancement of environmental protective measures and operating efficiency. Today's architecture could avoid the 460 ft. long driveway aprons at the front and rear and would present a more neighborhood-friendly profile.

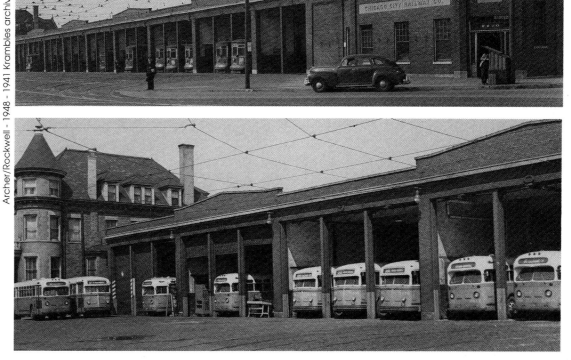

Archer/Rockwell - 1948 - 1941 Krambles archive

In CSL days, conversion of the 87th Street route to one-man operation in 1923 prolonged the use of streetcars there. After WWII, light riding, the need for through routing with a short bus extension, and deteriorated track mandated the change to buses on May 27, 1951. Within a couple of years the county programmed street resurfacing and later the wooden line poles and old-fashioned street lighting was also replaced.

Stony Island/87th - 6-22-53 - Krambles archive

47th/Lake Shore Drive - 12-27-47 - T. H. Desnoyers

The last extension of Chicago's streetcar network occurred in December 1937, when the 47th Street line was extended a short distance eastward under the Illinois Central Railroad tracks into Burnham Park. Streetcars on this line were replaced by buses on April 15, 1951.

Car 6091, built by Brill in 1914, was part of the first group of cars delivered to the unified management of Chicago Surface Lines. With this series began the practice that continued into PCC days of numbering cars bought for the account of the Chicago Railways distinctly from those for Chicago City Railway, in this case, 3000-3091 and 6000-6137, respectively.

Yerkes' normal practice in enlarging his network of car lines was to organize a new company under a separate charter for each extension, and then to operate the completed line in close connection with the parent. He claimed that this expedited service improvement, but it brought on Yerkes the image of a monopolist, a view exacerbated by his aggressive drive to blanket the city with car lines.

Suspicions and poor public relations generated in this period contributed to the creation of regulatory agencies and made it difficult for private management to establish credibility in later years when seeking relief from inadequate fares and service burdens.

Streetcar electrification continued throughout the city and there was constant improvement in equipment and operation. Finally, in 1906 the last of the cable lines (and also the last remaining horse car line) were converted to electric traction.

Under an Ordinance of February 1, 1914, all streetcar properties in Chicago were unified under Chicago Surface Lines management, streetcar operations outside the city limits having already been transferred to separate companies. In the process CSL became the largest surface transportation system in the United States under a single management.

Expansion of the streetcar system continued over the years until the early 1930s, mainly by extension of existing routes, although a few new ones were built. A substantial extension into "new" territory, on Western Avenue from 95th Street to 111th Street, was opened on November 8, 1931. Extensions eastward were made in the summer of 1933 to the Cermak and Roosevelt car lines to reach the Century of Progress Exhibition (World's Fair) on the lake front.

There were also some abandonments of streetcar services under CSL's period of administration. Most of these were replaced with motor buses or trolley buses. This type of mode conversion began in 1927 and was in progress at the time of CTA takeover. In a few cases there was no replacement service, there being alternate routes reasonably nearby, for example, on Elm and Franklin, from which cars were removed as far back as 1912. CSL operation on the interurban routes through to Hammond and East Chicago in Indiana was cut back to the city limit/state line when the independent Chicago & Calumet District Transit Company changed from streetcars to buses in 1940.

CSL placed its first five motor buses in service on a new route, Diversey, on August 11, 1927. Its first electric trolley buses were introduced on April 17, 1930, by coincidence on the same route extended. These installations marked the beginning of a thirty year decline of streetcar operations in Chicago.

Nevertheless, in the early 1930s, CSL participated with the streetcar operators of several

Undeveloped rural land on the southwest side was chosen for Ford and Dodge defense plants in World War II. They were linked by CSL bus routes on Pulaski and Cicero as well as a new rush hour shuttle service to 79th/Western begun on October 29, 1942. Even then the crowded parking lot in the background was a harbinger of problems ahead for transit. On a happier note, one of the wartime-built White buses like 3417 is preserved at the Illinois Railway Museum.

1944 - Krambles archive

other cities in the development of an entirely new design of streetcar, the so-called PCC car, incorporating many innovative ideas in body, chassis and propulsion design. Ultimately, 683 such cars were used on the surface lines, of which 570 were rebodied to continue further service as rapid transit trains. At press time, a few continued to ply the Evanston and Skokie Swift routes or be used in work train service.

Trolley buses, first used by CSL to extend service without having to do track construction, were also used in the conversion from streetcars. Over time, there were a total of twenty trolley bus routes. At their peak, there were seventeen lines using about 700 buses. They ultimately proved more expensive and less flexible than motor buses given the overwhelming street traffic congestion conditions in the 1950s after auto manufacturing resumed following World War II. The Kimball route in 1937 became Chicago's first example of replacing trolley buses with motor buses.

Chicago Rapid Transit Company

The earliest future component of the present CTA rail system was the Chicago & Evanston Railroad, tracing back to February 16, 1861, when the Chicago & Evanston R. R. was chartered. It is said to have begun operation of steam passenger trains between Calvary Cemetery (north of Howard Street) and Larrabee Street on May 1, 1885. However, its tracks didn't become part of today's Evanston route until 1908.

The first segment purpose-built for rapid transit service was the *Chicago & South Side Rapid Transit Railroad Company* between temporary terminals at Congress Street and 39th Street, using steam locomotives pulling light wooden coaches.

It became known as the "Alley 'L'" because, setting a precedent for an elevated railroad, it had purchased its own right-of-way, mostly along the back of lots adjacent to the public alley. The first train to demonstrate the line was one of six cars operated on Friday, May 28, and it is reported to have run the four mile line in 10 minutes, with 300 VIP guests. At 39th, they examined the structure while a lunch car was taken on from which a snack was served on the return trip. Coming back all the stops were made and the guests were given time to examine the stations. The trip was a perfect success. Everything worked satisfactorily and all were pleased with the action of the train and locomotives. During the following week, in preparation for opening day "trains were run on exact schedule time, but without passengers, to break in the gatemen and engineers."

Revenue service commenced on Monday, June 6, 1892. It was extended to the grounds of the World's Columbian Exposition in Jackson Park on May 12, 1893. The route was electrified in 1898, becoming the world's first to introduce *multiple-unit* control. Until then, electric trains of more than one car typically had required a motor car as a locomotive at the front end

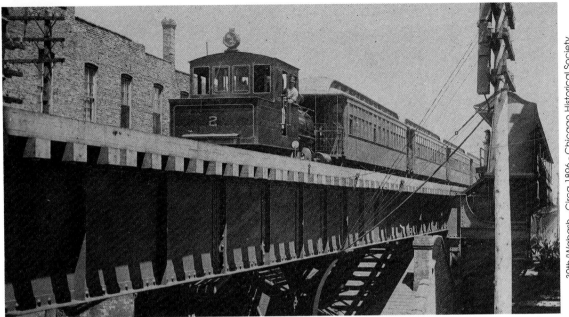

39th/Wabash - Circa 1896 - Chicago Historical Society

This is the way the initial stretch of the South Side Rapid Transit was built: nearly tangent and level, with elaborate brick station buildings tucked underneath. To build a much-needed third track over the adjacent alley, the existing tracks were raised enough to allow a mezzanine fare collecting area to be suspended below and the new track, along with its platform was built at a matching height alongside. The end product was a bit of a roller coaster, with twists to follow the existing alley. The resulting contour can be seen today. This is sometimes incorrectly presumed to have been done to provide momentum grades for the trains.

South from Washington/Market - Circa 1893 - Chicago Historical Society

The original downtown terminal of the Lake Street Elevated, a stub-end layout opened in October 1893, was at Madison/Market, reached by what wound up to be a branch used for afternoon rush hour departures. Use of steam locomotives (and locomotive cars after electrification in 1896) required a center relay track behind the camera. The branch to this stub was abandoned in April 1948 and the two-level Wacker Drive now occupies the route.

Through-routing of trains of the Northwestern Elevated Railroad to the South Side after 1913 is evidenced by the South Side destination sign displayed on Jewett-built car 1753. Smoking, once permitted only in cars so designated, was forbidden by city ordinance after October 14, 1918 as a result of the terrible influenza epidemic of that year. Car 1753 underwent an unusual amount of change over its life. In 1928 or 1929 its Hedley trucks were changed to Baldwin and its propulsion package was upgraded from GE-55 motors with type M control to W-567 with HLF control, identical to the newest steel cars. In 1957, after 51 years in passenger service, it became work motor S-332. It was again rebuilt to a derrick in the mid-1960s and in the 1970s it was one of the first air-braked cars to be modified for hauling as a trailer between cab signal-equipped 6000-class all-electric motor cars.

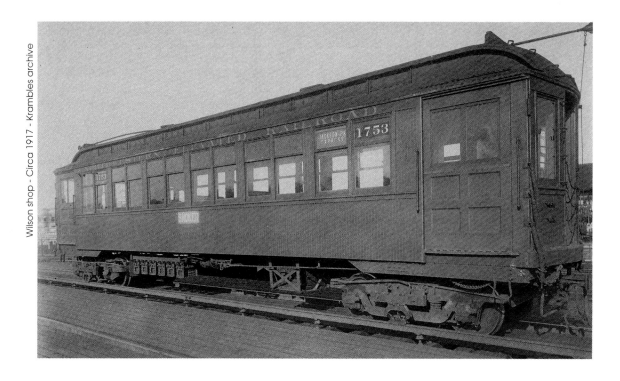

Wilson shop - Circa 1917 - Krambles archive

With its own electrification not yet complete, a South Side Rapid Transit train was hauled by steam through the 'L' Loop to the Metropolitan where it could be test operated on live third rail. On April 15, 1898 the first car ran electrically on the South Side and by July 27th all steam powered service had ended.

63rd/Cottage Grove - 4-17-1898 - Krambles archive

Central/Evanston station - 9-19-28 - Krambles archive

The original portions of the North Side 'L' were extended over ground-level tracks of the Chicago Milwaukee & St. Paul (steam) Railway. They were converted to the present elevated alignment in three steps involving difficult staging with unusual combinations of low and high, electric and steam tracks. Note gauntlet on nearest track, to permit wide freight cars to pass high platform. Service was maintained throughout the reconstruction, with special protective measures taken, such as enclosing stairways to keep passengers from coming in contact with the undercar equipment of passing trains, especially, of course, the third-rail shoes.

pulling the other cars as trailers.[2]

In 1905, to permit addition of an express track, the City allowed the use of air rights over the alley in return for the railway rebuilding its station fare control areas from ground level to mezzanine and paving beneath the two older tracks to create a wide alley. The Englewood, Normal Park, Kenwood and Stock Yards branches augmented the South Side system in the period 1907–1908.

The first operable segment of the *Lake Street Elevated Railroad Company* was opened on November 6, 1893, using steam locomotives. Electrification followed here in 1896. The *Metropolitan West Side Elevated Railroad Company* (Met) was the first rapid transit route in Chicago to be electrically operated, commencing service on May 6, 1895. Extensions and branches to the west were opened at intervals from 1902 to 1930.

The *Union Elevated Railroad Company*, the 'L' Loop, began use in 1897 as a terminal, the inner track operating counterclockwise serving South Side and Met trains and the outer track running clockwise for Lake Street trains. There were separate berths for each company so that each could collect its own fares.

The twentieth century brought into service the *Northwestern Elevated Railroad Company*, electrified from its opening on May 31, 1900. It was routed clockwise around the Loop, sharing the outer track with Lake Street trains. Both lines were under Yerkes' control, as was the Loop company itself.

The Northwestern was late in completion. It had pledged to forfeit a bond of $100,000 to the City if the road was not in service by January 1. From 400 to 700 men had worked day and night in rough winter conditions to meet the goal, but between Chicago Avenue and Lake Street it was only possible to complete one track so it was planned to run only one train a day until the double track could be completed.

The daily trip of January 2, 1900 was a source of excitement to the 25 passengers and numerous spectators. Starting from Wilson Avenue, the train was met at Lincoln Avenue by four policemen who arrested the crew and took them to the police station. However, there was another motorman aboard the train who took the controls and proceeded. At Lake Street there were 50 policemen on the track with

---

[2] The new concept, developed by traction pioneer Frank Julian Sprague who had developed the successful Richmond trolley, made two or more motor cars work in synchronization when coupled into a train. The ability of one car to pull trailers was then no longer a factor limiting train length.

In pre-subway evening rush hours overloaded station platforms around the Loop were endemic. There were no loudspeakers in those days, but a clerk in the booth on the overhead bridge followed the schedules and rang loud alarm bells to urge passengers and crews to hustle so that the next train, already waiting at the end of the platform, could come up. Mornings, when most passengers were *alighting* in the Loop, two trains were often berthed in a station at once, with resulting confusion of folks trying to anticipate which way to run to catch the train they wanted to board.

Adams/Wabash - 7-2-25 - Krambles archive

The lattice-work steel structure over 63rd Street was unlike any elsewhere in Chicago. It has a more-than-50-year history of costly repair. A structural engineer who studied it believes the design was probably chosen because it could be fabricated and erected in just four months to meet the opening date for the 1893 World's Columbian Exposition. The Fair terminal was actually a few hundred feet east (to the right) of the ugly duckling Jackson Park station, shown here, that served until March 1982.

63rd/Stony Island -- 8-1-46 - CTA

Latrobe/Lake - 1949 - G. Krambles

In the Lake Street corridor three transit modes come together for a mile between Austin, the Chicago-Oak Park city limits, and Laramie. A train of a 1906 Jewett-built and a 1903 St. Louis-built 'L' car, then operating from overhead trolley at this point, overtakes 1923 Brill-built streetcar while in the background a Chicago & North Western freight rolls by.
Today, new structure takes the rapid transit onto C&NW's elevation, C&NW tracks are a bit farther north and buses replace street cars.

Car licensing was a franchise ordinance requirement that died with the acquisition of CRT by CTA. Car 73, a "small Sprague", was last used in revenue service in 1930.

Krambles archive

orders to stop the train, but, instead of slowing down, the motorman increased speed and the police had to scatter as best they could. However, they then blocked the Northwestern track with piles of timbers to stop the return of the train, but it was sent west on Lake Street to Market Street, where it backed into the stub to be stored. Here the police finally caught up again and some of them had to spend a cold night in the train to make sure it couldn't return. The city claimed the company had defaulted on its ordinance obligation, but ultimately a court ruled for the company.

The Loop proved most difficult to operate. In rush hours, not counting any delay waiting in line for a "window" to enter, it required as much as 20 minutes for a train to circle the Loop, almost exactly two miles around. It would have taken much longer had there been signals to enforce the safe braking distance between trains, but they ran "on sight" in those years and could follow one another separated by only a few feet, depending on the skill and judgment of motormen to avoid collision, as is done in automobile driving even today.

In 1911, the owning corporations became affiliated for operational purposes as the *Chicago Elevated Railways Collateral Trust.*

On November 3, 1913, through routing of some service between the north and south sides of the city was introduced to increase the capacity of Loop trackage. The Lake Street and Northwestern, which had originally operated with left-hand traffic outside of the Loop, were changed to right-hand at that time. Both tracks of the Loop were changed to counter-clockwise operation and trains began to be routed in the basic pattern that was still in effect when CTA took over. Intercompany free transfer also was introduced and fare collection was consolidated.

In May 1934 a conflagration in the Union Stock Yards consumed everything in its path, including 'L' structure and a 2-car train trapped west of Halsted station. Restoration of service, delayed by the issue of who should pay—owner Chicago Junction Railway, operator CRT or the packing companies whose poorly protected stockpens formed the torch, resumed January 16, 1935. But more than a generation ago, the yards became redundant due to changes in the packing industry. 'L' service ended October 7, 1957.

To improve its share of the Century of Progress Exposition traffic, CRT was able to get temporary permits to operate shuttle buses from State/Van Buren and Cermak 'L' stations. Shown here is the Cermak terminal (top) and the staging area at the fair (bottom). Some of the buses used were loaners from the North Shore Line.

Extensions were made to Ravenswood, Evanston and Wilmette in the period from 1907 to 1912, to Niles Center (later renamed Skokie) in 1925, and to Westchester in 1926 and 1930.

On January 9, 1924, the ***Chicago Rapid Transit Company*** was incorporated to merge the former Northwestern, South Side, Metropolitan and Lake Street companies. This yielded some administrative economies.

In 1939 the City of Chicago and the U.S. Department of the Interior, Public Works Administration, joined to finance the construction of two subways through the central business district. On October 17, 1943, the State-Division-Clybourn subway was opened. The severity of congestion of train capacity on the Loop 'L' was finally relieved as through Ravenswood-Englewood and Howard-Jackson Park services were diverted through the subway. Completion of the Milwaukee-Dearborn-Congress subway had to be suspended until after World War II.

Halsted/Stock Yards - 5-19-34 - Krambles archive

August 1934 - Krambles archive

In two orders of 1916-1917 St. Louis Car Company supplied fifty double-deck bodies to Chicago Motor Bus Company. Riding on a home-built separable tractor unit with front-wheel drive and hard tires, this design was expected to reduce down time, but instead proved nearly three times as expensive to maintain as a conventional bus. The bus shown at right carried tractor number 23 and body number 114.

Chicago Motor Coach Company

The first motor buses used in mass transit in Chicago were operated by the *Chicago Motor Bus Company* on March 25, 1917. Its successor, *Chicago Motor Coach Company* (CMC), in 1922 fielded a minuscule fleet of only 63 buses.

It was a wholly owned subsidiary of *The Omnibus Corporation,* a firm also interested in motor bus operations in New York, St. Louis and elsewhere. CMC's great period of expansion extended from 1922 to 1927, by which time it had added 29 routes, was operating 134 street-miles with 432 buses and employed about 1,800 people altogether.

CMC ran mostly over boulevards which were under the jurisdiction, not of the City, but of the then-existing three Park Districts of Chicago. This became the source of the slogan, *"The Boulevard Route."*

The Boulevard Route operated profitably throughout the 1940s, providing good service during busy periods, but almost none at times of lighter riding such as owl periods, when prospective riders were left to find their way to paralleling CSL or CRT services nearby.

The CTA enabling Act and Ordinance of 1945 foresaw and intended that CMC should be integrated with CSL and CRT, but conditions at the time prevented immediate acquisition. For one thing, CMC was then still earning a return for its investors, so helping to save it was not urgent. More importantly, the credibility of investing in transit, even be it a public authority, was still in doubt, as evidenced by the fact that it took practically two years to sell only $105 million in revenue bonds. Prudently taking first things first, the initial bond issue was kept to a minimum. However, by 1952 CTA had established the credit to issue a further $23 million issue, of which $16.4 million was used to buy CMC.

Montrose-Wilmette line (CMStP&PRR)

The portion of the north side 'L' between Montrose (Graceland) and Wilmette (Llewellyn Park) had been built in the 1870s by a predecessor of the *Chicago Milwaukee St. Paul & Pacific (steam) Railroad*.[3]

Under a leasehold agreement it had been rebuilt in 1908 for operation of Northwestern Elevated Railroad trains to Central Street, Evanston, and on to Wilmette in 1912. Beginning in 1920 operation of freight trains over this trackage by Northwestern using electric locomotives superseded the service previously run by the steam railroad and the tracks were also elevated. In 1953, CTA acquired full title for $7 million, obtained from a final revenue bond issue in that amount. By 1973 coal traffic had atrophied and freight service ended.

Other acquisitions

The possibility of further system expansion by purchase was considered by the CTA Board, in an extensive study made in 1948 of the west suburban bus systems then being operated by the *Chicago & West Towns, Leyden Motor Coach,* and *Bluebird Coach* companies.

However, under the Trust Agreement securing the revenue bonds CTA could not issue the additional bonds that would have been needed unless a study could confirm that revenues from such acquisition would equal or exceed 200% of the total of the added operating expenses, plus the principal, sinking fund and interest charges that would result. Unfortunately it was already too late in the skid of suburban bus service viability and no new operating plan could be devised to satisfy the stipulated terms.

All CTA revenue bonds are now retired, so this constraint would no longer apply, but some alternative funding mechanism would be needed. A broader solution to the problems of outer suburban bus service was undertaken a quarter century later when the **Regional Transportation Authority** (RTA) was born. Now, a separate entity, ***Pace***, has become the suburban bus operating authority.

◊ ◊ ◊ ◊

---

[3] Interestingly, the Chicago & Milwaukee Electric Railway, building its interurban line southward from Waukegan, in 1899 cut a deal with the Chicago Milwaukee & St. Paul Railway to use the latter's track between Llewellyn Park (today's Wilmette) and downtown Evanston. Thus, Northwestern Elevated and C&ME overlapped at an early date, both as tenants on the same railroad.

This was S-105, one of two Baldwin-Westinghouse locomotives completed in August 1920 to take over freight service north of Wilson Avenue on behalf of the Milwaukee Road (CMSP&PRR). Movement of coal for residential and commercial heating was always the predominant activity. As the end of freight service approached, the Lill Coal Company's yard at Berwyn station was the only remaining customer; oil and gas had become the fuels of choice in cities. Between Irving Park and Howard, freights operated only on the west track; in Evanston, they used the east track. Gauntlets provided the clearances needed for third rail and wide cars. For a period beginning in June 1942, the two locomotives were rented to the North Shore Line Mondays through Saturdays 0930 - 1830. *(See also page 110.)*

Dempster, Niles Center - Circa 1926 - Krambles archive

Chapter 4 - Policy issues 1945-1970

Cermak/Cottage Grove - 6-6-52 - T. H. Desnoyers

Chapter 1 told about the formation of CTA as an independent municipal corporation to provide transit service in Chicago. Its predecessors, under *private ownership* with *public regulation,* had repeatedly failed to generate sufficient revenue from fares to meet the combined cost of operation, plant renewal, modernization and expansion needed to match the public service demands imposed by the growth of the city and its suburbs.

The State delegated responsibility to the CTA Board for determining fares, routes, kinds and levels of transit service in most of Cook County. To acquire the assets of the then-existing privately-owned streetcar, bus and rapid transit systems, revenue bonds were sold, to be

To spread out the expense and work load involved in bus conversion, some streetcars were modified for one-man operation. Here is a 1936 PCC as adapted for such use on Route 4 - Cottage Grove. But there were many problems getting political and union acceptance of one-man cars. Other approaches were tried, such as weekend buses with streetcars remaining on weekdays. In the end, direct conversion to motor or trolley bus proved the best method.

Parkside, Lake - August 1950 - G. Krambles

In the absence of automatic block signals for rear-end protection of trains, CRT used "spacing boards" as a guide to motormen in maintaining safe following distances (don't pass one until the next is in plain view). Here spacing boards are seen in the devilstrip between tracks. For night visibility some had lights inset in the mast.
(See also photo on page 70.)

30

paid off from revenues of the combined operation. The bonds and system-generated revenues were the only sources of financing originally provided to cover all capital and operating costs of the property.

With these givens, CTA pursued a cooperative urban and suburban approach to providing a metropolitan transit service. In the Chicago region, the Metropolitan Water Reclamation [formerly *Sanitary*] District is the only municipal corporation that provided *regional* service before CTA.

CTA was able to replace streetcars with a first generation of small buses, and when those wore out, a bigger and better second generation was provided. The change to buses made it possible to avoid the repair or replacement of life-expired streetcar track and pavement; municipalities promptly covered them with blacktop. After an innovative approach to financing was conceived by then-Chairman Ralph Budd (retired Chicago Burlington &

Washington/State - 1949 - Krambles archive

Exterior car washing of steel cars was suspended for a while in the late 1940s, to test whether the grime and grease of subway operation would be enough protection to ward off corrosion. It wasn't, and soon a massive cleanup and painting campaign not only revived the cars but caught up to eight to ten years of wartime wear and tear that had worn off the sheen of station finishes in tunnels.

Rationalizations instituted early on eliminated the services that operated from stub terminals outside the Loop: west from Wells Terminal, north from North Water, and south from Old Congress. Shown here is the last named, the original South Side terminal, closed 7/31/49. These p.m.-rush-only trips had atrophied in usefulness with the coming of subways.

Holden Court/Congress - 7-22-49 - T. H. Desnoyers

Quincy R. R. President), combining buses with cars in a single equipment trust issue, a start was made in replacing obsolete rapid transit cars. Shops, garages, substations and structures each received a modest share of modernization. All this caused a significant gain in system productivity as well as the improving transit's public image.

CTA had acquired the world's only rapid transit most of which was without an automatic block signal system and depended on careful "on sight" driving to avoid rear end collisions. It was able to make a quantum leap forward in passenger safety in 1965 by the introduction on the Lake route of an innovative system of cab signals, utilizing audio frequency track circuits and providing continuous speed control.

69th/Parnell, south terminal - 8-7-46 - CTA

Harvard/Englewood station - 4-1-50
T. H. Desnoyers

The Normal Park branch, 0.93 miles from Harvard to 69th, had two intermediate stops. For years it was served by a car added to (or cut from) Englewood trains at Stewart Junction, just west of Harvard. Under the North-South A & B plan of 1949 service shuttled, using old cars and turning through spring switches. MCERA Bill Janssen recalls working a run that made 28 round-trips on the line! The branch closed without replacement January 29, 1954.

103rd/Vincennes - Circa 1950 - Chicago Aerial Survey

Service extension and route rationalization kept pace with changing land use. A multi-agency approach was adopted for costly changes to the rail system, with participation of numerous federal agencies, the State of Illinois, the County of Cook, the City of Chicago, and the Villages of Skokie, Oak Park and Forest Park.

Examples began with the Dearborn Street (West-Northwest) subway, the first section of which opened in 1951, and by 1970 also included the relocation of the Garfield 'L' into the Eisenhower Expressway, elevation of the west part of the Lake 'L' out of street running, and in the Skokie Swift, Englewood, Dan Ryan, and Kennedy extensions.

As Chicago's development expanded garages were needed closer to the extremities of the routes. Beverly garage, at 103rd/Vincennes, was the first new one, opening on December 4, 1949. Outgrown and outmoded by two decades of change, It is now used by utility and buildings/grounds groups, having been replaced by 103rd/Stony Island garage.

Although much tunneling and station structural work of the Milwaukee-Dearborn subway was concurrent with that on State Street, wartime material shortages held up completion at each end, station finish, track, signal, other installations and new cars. Finally, in February 1951 the Logan Square to LaSalle/Congress segment was opened. Here on a VIP preview of the new line was only recently delivered car 6101. Facing right at the foot of the stairs is George DeMent, then Chicago's Commissioner of Subways & Superhighways, later to be CTA Chairman.

Washington/Dearborn station - 2-23-51 - Krambles archive

There were gains in cost-effectiveness through the conversion of streetcar to bus service, thus avoiding two-person crews and heavy maintenance costs for tracks, paving and power facilities, a formidable portion of which were life-expired, their unreliability interrupting the reliability of service.

Remote door controls and a few one-person trains were introduced on rapid transit to cut crewing size. Station platforms were lengthened to allow the use of all side doors and longer trains. Duplication of paralleling surface and rapid transit routes was reduced. Very long bus routes, so difficult to maintain regular, dependable services on, were reconfigured into more manageable segments and routed to feed rapid transit stations. Several new garages were built to replace hard-to-work-in streetcar barns. Other obsolete facilities were phased out.

Parkside/Lake - 1962 - G. Krambles

The Lake rail route ran at street grade adjacent to the Chicago & North Western Railway between Long Avenue in Chicago through Oak Park until October 1962, when CTA joined with the Village, the City, the County, the State and the U. S. Bureau of Public Roads in sharing the $4 million cost of relocating the 'L' to the C&NW elevation. Coincidentally, improvements in service, including gradual acquisition of new cars, meant frequent changes in types of rolling stock used there. MCERA Glenn Andersen recalls busy years spent reinstructing car repairmen as the various cars were cascaded through the system, each change bringing a different variety of equipment detail to be mastered.

22nd/Mannheim station - 12-1951 - B. L. Stone

Terminal station of the little-used Westchester branch as it appeared 22 years after start of service and just before trains were replaced by buses. The all-too-little potential for walk-in ridership is evident.

Early on, CTA policy was to integrate bus and rail modes so that each type of service would do what it best does. Adaptation to constantly changing land use makes ongoing review of route and station locations imperative if a transit system is to satisfy public convenience and necessity with minimal subsidy. Applying this approach to the rail lines, more than 40 new rapid transit stations were opened while over 100 little used old ones were eliminated. As an example, Grand station, one of those closed, was within a few hundred feet of Merchandise Mart, a station now completely rebuilt to modern standards. Last patronized by only a few dozen, Grand's removal speeded up travel for thousands of rush hour riders who had been inconvenienced by the stop.

Grand/Ravenswood station - August 1970 - G. Krambles

Yet transit seemed to be stuck in a market ever-shrinking in volume yet ever growing geographically. Expansion of the heavily-subsidized urban street and expressway network continued to entice riders away from CTA. Automobile population continued to grow; parking spaces replaced buildings that had sheltered thousands of work stations. Competing with transit became a growth industry even in the face of oil shortages and environmental disasters. Dispersion of residence and employment into the rural periphery accelerated. So transit's market share kept shrinking while the area it was expected to serve expanded and the average distance of each trip got longer.

In the meantime, inflation pulled wages upward faster than technological change could be implemented to improve productivity. The prices of fuel, power, materials and services climbed. To renew worn-out equipment, everything now cost more than had what it replaced.

Unfortunately, CTA's capital was in effect frozen at the $135 million level of the revenue bond issues which had paid for acquisition of the properties and startup. Even though shrinking traffic required a little less equipment each year, those resources had barely paid for basic replacements in never-to-return low cost years. As a result, by 1970, the transit dilemma had again become critical.

◊ ◊ ◊ ◊

Chicago's 25-year experience with propane was unique in the bus industry that was at the time generally switching from gasoline to diesel as a fuel. The 8700-series of 1963, totalling 150 buses, were the last propane-powered coaches to be purchased.
In service they proved a little sluggish. Inability to find a qualified bus builder offering a more powerful engine was one reason for the end of Chicago's propane era in 1975.

Chapter 5 - Policy issues 1971-1991

The first half of CTA's life had seen it develop as a self-supporting service, paid for entirely by user charges, without funding assistance by tax or other subsidy.

But unforeseen circumstances made it clear by 1970 that if this approach were to continue, CTA would rapidly self-destruct. In the attempt to keep pace with rising operating and capital costs, basic fares had been increased from 1947's 10¢ (12¢ on rapid transit) to 45¢, but originating fares per year had skidded from 888 million to 296 million on surface lines and from 146 million to 106 million on rapid transit.

Each fare increase was paralleled by a decline in ridership. Each drop in ridership was followed by an incremental decrease in the amount of service offered, although it was impractical to make precisely corresponding reductions in seat-miles operated by running shorter distances or less frequently.

Population of the city itself had dropped from 3.6 to 3.4 million, while automobile registrations rose from 554 to 913 thousand. Average rides taken per capita in a year went from 315 to 187. Riding pattern shifts to outlying areas of the city and contiguous suburbs served by CTA are more difficult to quantify, but there is no question that this effect raised the average trip length and so increased the cost to produce it.

A comprehensive limited-access highway system (known in Chicago first as *superhighways* and now as *expressways,* but elsewhere mostly as *freeways*) was being built into and through the heart of the city. Generally 90% of its cost was paid for with federal funding under the Interstate Highway program. While the resulting network did greatly relieve the paralleling local surface streets, congestion was actually worsened on key streets feeding the expressways. But the use of private automobiles was definitely facilitated and, on a long term basis, dependence on foreign-supplied fossil fuels was increased. Land use was drastically altered with more area used for parking. Strip commercial development along arteries one-half mile and one mile apart (exactly those predominantly used by transit) was superseded by shopping centers occupying whole square blocks of land at scattered sites. The most attractive of these were in suburban areas, away from city transit, diverting merchants and customers from downtown and neighborhood stores. Vacant shops left behind led to blight and a vicious cycle of deterioration. Together these impacts seemed to focus on public mass transit, raising its costs and lessening its revenues.

The CTA Board took the problem to its "founding fathers," the State of Illinois General Assembly. Here it became apparent that urban transit was not just a problem local to Chicago, but it was also the experience throughout large and small urban areas throughout the state. Some downstate communities had already lost all transit service. All others were in crucial stages of atrophy. Development of new approaches had to begin.

To supersede its standing Highway Department, the State created a Department of Transportation with several programs in it to provide direct financial assistance to publicly-owned transit operators. Although modest at first,

Comprehensive reconfiguration of Howard yard was one element of preparation for the February 1993 linking of Howard service to the Dan Ryan branch (instead of to the Englewood and Jackson Park lines, which were simultaneously linked to the Lake line, Dan Ryan's former partner). Included at Howard is a new inspection shop (replacing the 1900-vintage Wilson facility) and a modern solid-state microprocessor-controlled interlocking plant. The difficult project involved many teething problems, testing riders' and employees' patience.

Howard/North-South, Evanston, Skokie Swift yard - 10-23-91 and 11-21-91 - A. Peterson

this was of great importance in staving off the collapse of service throughout the state, including of course the Bi-State agency which serves the St. Louis–East St. Louis area.

The special problem was the metropolitan Chicago region spreading over six counties of Northeastern Illinois: Cook, DuPage, Kane, Lake, McHenry and Will. To serve this area, there were, in addition to CTA, several railroads and about a dozen bus lines operating suburban passenger services. Solution here required taking these over from their private owners and to do this, the Illinois General Assembly created by statute an overall agency, the **Regional Transportation Authority** (RTA), which was approved by referendum in the six counties in March, 1974.

RTA's mission was to coordinate and provide a consistent level of financial support to the diverse transportation system. It began as an operating as well as administrative agency, providing day-to-day bus and rail service through grants to CTA and the several suburban transit districts that had been organized to try to stabilize bus and railroad operations.

RTA created one subsidiary (today's **Metra**) to acquire, operate, or contract for operation of commuter trains, and another subsidiary (today's **Pace**) to do likewise for suburban buses. RTA also supports the South Shore Line, extending into Indiana, where its funding responsibilities fall to the Northern Indiana Commuter Transportation District, which also operates it.

However, to relieve political and financial problems, RTA was reorganized by the Illinois General Assembly in 1983 to become a funding and oversight entity. Responsibility for management and operation was shifted to three so-called Service Boards: **Metra** for commuter rail, **Pace** for suburban bus. **CTA** continues for bus and rapid transit in Chicago.

In the 1980s the City of Chicago completed construction of extensions from Jefferson Park to O'Hare Airport.and CTA began operating them. On February 21, 1993, a major reconfiguration changed the through routing of North-South and West-South services, aligning Howard to Dan Ryan and Lake to Englewood-Jackson Park. Also nearing completion was the new southwest route to Midway Airport.

RTA developed a strategic plan identifying and quantifying funding needs. Anticipating decreasing federal funding and growing dependency upon local resources, legislation in 1989 enabled issuance of $1 billion of bond debt by RTA to expedite rehabilitation of the system. Unfortunately, a general business recession withered sales tax revenues, once again frustrating transit development.

◊ ◊ ◊ ◊

When CTA took it over, CSL had about 3700 cars and buses, but with the exception of the very newest, the rolling stock was exemplified by this car, built in 1906 by J. G. Brill Company of Philadelphia to a Chicago City Railway design—by then worn, but serviceable. Similar cars from a variety of builders served all over the system until the last were retired in 1954. Some of the 683 PCC type cars, then considered quite modern, worked on until 1958 and even then were remanufactured into rapid transit units, a few of which remain in 1993.

At the time of their introduction, PCC streetcars represented a major breakthrough in comfort, speed and noise reduction. Within a year of their introduction, the eighty-three 1936 PCCs were credited with a 12% increase in ridership on route 20-Madison with its 5th Avenue branch. Significant in the good result were strong intersecting feeder lines, like the pioneering electric trolley bus route on Central Avenue. Part of the gain was won competitively from paralleling rapid transit, then a predominantly local or low speed express service.

1937 - CSL

Chapter 6 - Streetcars and buses

As of January 1, 1945, the equipment inherited by CTA for operation of the surface part of its system included about 3,269 streetcars, 152 trolley buses, 259 gasoline buses and more than 300 utility (work) cars—a truly huge fleet. To pin down exact numbers as of October 1, 1947, CTA's "D" day, is to chase a moving target; obsolete cars were being scrapped at the rate of up to 55 a week, while new streetcars, motor buses and electric trolley buses were in various stages of delivery from the builders.[1]

Streetcars

Since Chicago Surface Lines was an *operating* entity, ownership of its rolling stock actually resided with four of the five underlying companies. Penciled notes in an old file indicate that on October 1, 1947, the passenger car fleet included 1,929 ex-*Chicago Railways* cars on hand, 1,125 ex-*Chicago City Railway,* 19 ex-*Calumet & South Chicago,* and 27 ex-*Southern Street Railway.* Of the new streetcars on order with CTA approval, 80 had been received for the account of the *Chicago City Railway* and 124 for the *Chicago Railways*.

Among the old cars at CTA takeover, the most prevalent were the three groups of similar cars numbered 101–1100. There were 883 of these so-called "Big Pullmans, Stove-Bolts, and Small Pullmans" still in use. They dated back to 1908–1910. The next most numerous cars were the three similar Brill-built groups in series 5001–5600, dating from 1905 to 1908, of which 471 remained at CTA takeover.

Despite many variations between the individual groups, the older streetcars being used in 1947 conformed pretty much to these general characteristics:

Length	41' to 49'
Width	8'-6"
Height over roof	11'-10"
Truck centers	16' to 22'
Truck wheelbase	48" to 72"
Seats	40
Weight	22 to 27 tons
Number of motors	2 or 4
Total power/car	120 to 200 hp
Power supply	600 volts DC
Control, 225 cars	Single-end
All others	Double-end

[1] While CTA didn't take over operation until October 1, 1947, by agreement with the Federal District Court it was authorized to exercise all the powers of the Metropolitan Transit Authority Act from June 4, 1945, the date of the successful referendum, and immediately began to purchase modern streetcars and buses using cash from CSL's renewal funds.

Normal top street speed for these old cars was 28 mph, except for the so-called "sedans" of 1929, which had a free speed of about 35 mph and much greater acceleration.

Chicago Surface Lines had been an active participant in the Electric Railway Presidents' Conference Committee (PCC for short), a group representing the leading streetcar companies which had undertaken to research and develop an entirely new vehicle concept. CSL had tested two complete prototype cars as well as other developmental propulsion packages. It was one of the first railways to purchase a production lot of the PCC design with 83 cars acquired in 1936. As one of the initial steps in the CTA modernization program, with CTA and court concurrence orders were placed for 600 more PCCs of an improved design in 1946. Of these 103 were received in 1946, 199 in 1947 and 298 in 1948.

The principal characteristics of these PCC cars were:

Length	50'
Width	9'-0"
Height over roof	10'-3"
Truck centers	22'-9"
Truck wheelbase	6'-3"
Wheel diameter	25"
Seats	57
Weight	20 tons
Power	4 motors @ 55 hp
Free speed	41 mph
Lot cost	$15,133,914

After 1946 no more streetcars were ordered by CTA; in fact, hardly were the deliveries of that batch of cars completed than policy decisions had been made that would ultimately phase out all streetcar operations in favor of motor buses, trolley buses or rapid transit.

Electric Trolley Buses

The transition from streetcar to electrically powered trolley bus was a natural and obvious one for CSL to make and CTA to pursue. The technology was only incrementally different from the mature hardware of the streetcar, so there was little risk involved in the design, construction, operation and maintenance of the new mode.

The existing 600 volt DC power conversion and distribution system was easy to adapt to the new electric bus routes. As compared to gasoline

Kedzie barn -1935 - Krambles archive

For the introduction of PCC cars in 1936 CSL developed a unique pedestal-mounted fare register having a translucent top. Levers recorded cash, token and transfer full and half fares. With each recording a differently colored light illuminated the top and a distinctive bell tone sounded. Following the Peter Witt layout for two-man cars, the conductor's location was near the center of the car. With the 1946 PCC orders, the conductor was located at the rear of the car, which CSL felt improved fare control and reduced transfer abuse. Conventional fare registers were applied.

Opposite page - **After the 1945 referendum modernization was pushed forward with equipment orders initiated with CTA concurrence. By 1946, with takeover still a year away, CSL was receiving deliveries on orders for 600 PCCs similar to 4160 shown.**
Madison/Central Park - 1946 - Krambles archive

Early 1948 - Krambles archive

Posing on Central at Wabansia, off the route for which it is signed, St. Louis Car Company-built 385 was part of the first postwar order of trolley buses. A heavy vehicle, it had superior riding qualitiies. In common with other electric trolley buses of the period, it was equipped with sanders.

Diversey/Western - 2-13-49 - W. C. Janssen

Built as a demonstrator by Twin Coach in 1946 as a gasoline bus, 9763, the "Supertwin", was converted two years later to a trolley bus and finally was purchased in 1954. It was 104" wide and 47' long. As is shown here, the articulation permitted the rear body section to hinge vertically, but not horizontally. Today's artics permit both. The Supertwin saw regular service on the Diversey and Belmont routes. It is preserved at the Illinois Railway Museum.

motor buses, the vehicles were easier to drive, quieter and involved no offensive odors of fuel or exhaust. As compared to streetcars, no track was needed; just poles and wires for the power distribution, although these were a little more complex than those for streetcars.

In the years 1930-1973 during which electric trolley buses served Chicago there were innovative technological developments, for example, the development of dynamic braking to a stop. Before, dynamic faded at walking speed and air brakes took over. The change increased tire and brake life and provided smoother stopping.

Another interesting test was of the **Supertwin,** a 58-seat 47-foot, 3-axle, 21,510-pound giant, a precursor of today's high-capacity articulated buses. Actually, although the Supertwin was hinged in the middle, it could only flex on the horizontal axis, unlike modern articulated buses which bend freely both horizontally and vertically.

The general characteristics of CSL's first electric trolley buses, of 1930 vintage, as compared to the last, built in 1952, were:

Year delivered	1930	1952
Length	31'-7"	39'-10 5/8"
Width	91"	102"
Seats	40	49
Light weight	19330 lb	20340 lb
Motors	2 x 50 hp each	1 x 140 hp
Free speed	30 mph	40 mph

The chronology of the application of electric trolley buses in Chicago from their inception to phaseout is given in the following table:

Route	Miles to make a round trip	First day	Last day
Diversey	12.7	4-17-30	6-18-55
Central	16.2	6-08-30	1-17-70
Narragansett	7.1	6-29-30	1-31-53
Elston	5.5	7-31-30	1-21-51
Montrose	13.7	1-25-31	1-13-73
Belmont	19.4	5-30-31	1-13-73
Kimball	3.4	6-21-31	6-30-37
51 - 55 Street	14.2	6-20-48	6-20-59
Irving Park	14.6	11-07-48	1-13-73
North	16.4	7-03-49	3-23-73
Fullerton	12.7	12-04-49	1-22-73
Lawrence	12.6	4-01-51	12-13-69
Pulaski	21.7	9-16-51	3-24-73
Cicero	15.9	11-25-51	3-24-73
Grand	21.3	12-16-51	1-15-73
47 Street	12.7	3-12-52	6-20-65
Chicago	16.4	5-11-52	3-26-67
Armitage	13.1	2-01-53	10-15-66
Roosevelt	15.0	5-24-53	1-13-73
Kedzie-California	23.8	12-04-55	3-15-69

Table shows first day and last full day of revenue service by trolley bus. Some lines were phased into service, some were phased out, others were extended after startup date shown. In the case of Montrose, trolley bus service was suspended for a period of about one year from 4-24-47.

Modernization in the early 1950s included 500 Twin Coach model 52S2P propane buses and 349 Marmon-Herrington model TC48 electric trolley buses (an all-time industry record ETB order). All of these coaches had double-stream front doors to speed up loading. In those days operators had to make change and it helped to have a front platform that could swallow a big group of boarding passengers.

Motor buses

Chicago Motor Coach and predecessors

The *development* of motor buses in Chicago is primarily a CTA story, although *introduction* of buses in Chicago must be credited to the Chicago Motor Bus Company, which began service on its 9.5 mile Sheridan Road route on March 25, 1917. The earliest vehicles had 51-passenger double-deck bodies built by St. Louis Car Company and were powered with detachable front-wheel-drive "tractors" having 50-hp gasoline motors.

Although Chicago Motor Bus had been seeking the right to operate south side routes, the poor performance of their crude equipment helped the Circuit Court decide on January 8, 1918, to award them to the competing Chicago Stage Company, sponsored by New York interests who proposed a vehicle similar to those then operating on the Fifth Avenue Coach lines in Manhattan. A third contender was the Depot Motor Bus Lines, which began a line between Union Station and the State Street store of Carson Pirie Scott & Company in 1920.

Under the leadership of John A. Ritchie, the motorbus lines were brought together to form Chicago Motor Coach Company and they were expanded to cover more territory including the west side of the city.

Confrontation with the competing (and senior) Chicago Surface Lines, festering for years, became open when Motor Coach invaded the northwest side of the city in 1928 with five new routes totalling about 17 miles of streets. After many months of court hearings, Surface Lines won the rights and CMC had passed the pinnacle of its route structure.

All the while technological change progressed. Pneumatic tires replaced the solid ones in 1927. Double-deck buses had been on the system since 1917, but by 1927 enclosed upper decks were introduced. By 1933, there were 354 double-deckers and 192 single-deckers in the fleet and in the Century of Progress Exposition (World's Fair) that year, one of the exhibits was a new double-decker with engine in the rear, providing 73 seats in its 33 ft length. This was a prototype of the classic "Queen Mary" series, 100 of which were ordered in 1935.

The use of double-deck buses was constrained in Chicago, strangely enough, by the remarkably forward-looking policies the city had adopted in the 1890s, forcing railroads to grade separate by track elevation. Railroad center that

Wilson/Ravenswood - Circa 1935 - Krambles archive

CMC continued to refine double-deck bus design throughout the 1920s and 1930s. One-man operation eliminated the conductor, who previously made warning announcements and visually monitored the open upper deck while passing under low bridges. To address safety concerns, warning systems were tried, including electric signs and loudspeakers. Ultimately, for comfort as well as safety, a way was found to lower the floors and both decks were fully enclosed.

The high-point of double-deck design for Chicago was achieved in a 1934 Yellow Coach design which became known as the "Queen Mary". Bus 104 was the sample for CMC of this 72-passenger one-man design. A final forty buses of this type were received in 1937-1938 and were converted from gas to diesel in 1941. They were Chicago's last double-deckers, running until July 1950. With the integration into the unified system in 1952, duplicative numbers of Chicago Motor Coach Company's bus routes were avoided by adding *100* to their former numbers, for example CMC route *51-Sheridan Road* became CTA route *151-Sheridan Road*.

Devon/Sheridan - 1935 - Krambles archive

it was, Chicago was laced throughout with tracks and it was obvious early that commerce would be strangled without extensive amounts of grade separation.

However, much of the early work provided clearance between pavement and underside of structures ranging only 11 to 13 ft., much less than needed for a double-decker with comfortable headroom. Early double-deckers ducked the problem by leaving unenclosed the upper deck (which was practically useless in winter or on rainy days) and by leaving so little headroom on the lower deck that passengers had to stoop.

It was not impossible to touch the underside of certain viaducts while seated on the open upper deck. One extreme example: on Route 51, passing under the Surface Lines' trolley wires suspended (hardly a foot away, it seemed) from the Loop 'L' structure along Wabash Avenue. For this and other reasons the single deck configuration ultimately became standard for bus service in Chicago.

Diesel-powered buses with automatic hydraulic transmissions appeared first in 1939. To serve wartime traffic, the fleet was increased to a peak of 619.

By 1952, when Motor Coach was consolidated into CTA, its fleet consisted of 595 buses, nearly all diesels, operating on 23 routes out of five garages.

Chicago Surface Lines buses

Chicago Surface Lines' experience with motor buses began in 1927 with a modest five-bus fleet built by Twin Coach, each having dual six-cylinder 55-hp gasoline engines placed out of sight under the floor, an innovative concept at the time. The bodies resembled those of streetcars. They served the pioneer CSL bus line, on Diversey Avenue between Laramie and Crawford (now Pulaski Road), 1.5 miles.

Bus operation by CSL expanded slowly at first. Small lots of equipment were purchased from several competing suppliers, but chiefly from Twin Coach and White. On January 1, 1945, before renewal funds were made available to CTA, the CSL rostered 259 motorbuses. But already CSL had planned massive modernization involving abandonment of streetcar track and conversion to motor bus service and, to stanch the hemorrhaging of operating losses,

In the brief three years beginning in 1946, expansion of the bus system and conversion of streetcar routes required delivery of 1,100 buses, a quarter of which were model 798 supplied by White Motor Company, then an active bus maker. Bus 3598 was one of a 100-unit order of January 1948. They were among the first Whites to have automatic transmission.

Archer/Rockwell - 1948

Devon/Sheridan - 1946

Milwaukee/Foster/Central - Circa 1945
Three photos this page - Krambles archive

CMC's enclosed doubled-deck buses were similar to those of New York's affiliated Fifth Avenue Coach, but had the lower profile needed to pass under viaducts which seldom offered as much as 14'-6" clearance.

Trolley bus destined to Central-Lexington, gas bus to Lehigh-Tonty and Milwaukee streetcar going Downtown via North Western Station. Evolution of modes is illustrated by streetcars having reached here by 1914, trolley buses by 1930 and motorbuses in 1940. Today all three routes are operated with diesel motor buses.

46

Chicago Motor Coach, acquired by CTA late in 1952, was quickly merged operationally into what was then called the "Surface System." CMC's newest vehicles were in their original livery, like 635 shown here northbound at the Halsted/Englewood 'L' station. The Route 8 Halsted streetcar had been converted to bus on May 29, 1954. Actually, the 600s came in two lots of 50 buses each. The second lot introduced fluorescent lights to Chicago buses. The last of these ran revenue service in 1972.

63rd/Halsted - 12-7-56 - Krambles archive

the program was begun with full concurrence of the CTA Board and the Federal court.

CTA begins bus conversion

The rapidity of this action is demonstrated by the record of bus equipment purchases for the years 1945-1954:

Year	Trolley	Gas	Diesel	Propane
1945	-----	55	-----	-----
1946	-----	172	20	-----
1947	-----	271	10	-----
1948	210	272	100	-----
1949	-----	-----	-----	-----
1950	-----	-----	-----	31
1951	190	-----	-----	520
1952	159	-----	Δ----	-----
1953	-----	-----	-----	105
1954	2	-----	-----	* 295
Subtotal	561	770	Δ 130	Δ 951
Total, all buses				Δ 2412
Lot cost				$40,295,516

*Of a further order of 100 propane buses, 22 had been received by December 31, 1954, but were paid for in 1955.

Δ Not including 595 buses acquired with purchase of Chicago Motor Coach Company.

Propane fueled buses

Replacement of streetcars created an opportunity and necessity for accelerated improvement in bus technology of which CTA was acutely aware.

Clearly the small gas bus of prewar design was not going to be the answer, but what was? Maybe the trolley bus? Maybe the diesel: the war had brought much improvement to the design and power of the diesel engine. Hedging its bets, CTA sought to diversify its purchases. The introduction of the propane bus provided a remarkable innovation for CTA to latch onto.

The first liquefied propane gas (LPG) buses were tested on CTA in 1950 and the results were so pleasing that at their peak in 1963 there were 1,671 of them were in the fleet of 2,658 motorbuses. CTA's was the largest such fleet in the world, although about twenty other transit systems used the technology. Yet, by 1975 the CTA fleet was practically 100% diesel powered and the quarter century of propane bus predominance passed into history.[2]

In a brochure, *"Facts about LP-Gas"*, issued in 1957 when there were 1,301 propane buses in the fleet, CTA described its reasons for using propane and its experience with it:

"Propane is a completely odorless motor fuel. It burns without a trace of a smell. It is also a non-strategic fuel, and consequently should be abundantly available if national defense considerations should require the rationing of gasoline, and the fuel oils that power jet planes...

---

[2] The gasoline bus had been used in Chicago transit from 1917 to 1963. The electric trolley bus era ran from 1930 to 1973.

Washington/Central Pk - 1960 - CTA

In 1960-61 GMC got its first CTA order, a lot of 150 (including bus 111 shown), which together with a lot of 150 from Flxible, introduced the "New Look" design to the fleet. Actually, the style had come to Chicago in 1959 with equipment received by suburban carriers in the area.

Not evident in this view, Chicago Motor Coach's 500-600 series was unique, having a length of 43 ft. A production order of 100 buses was followed by the purchase of the GM demonstrator. All were retired by 1970.

Randolph/Michigan - 1952 - Krambles archive

"The fuel tank on a propane bus is constructed of heavy quarter-inch boiler-plate steel, capable of withstanding a pressure of 1,000 pounds per square inch, although actual working pressures are approximately 190 psi. In the event of a collision, a propane tank has over twenty times the resistance of an ordinary gasoline or diesel bus tank....

"Between the tank and the engine, the liquid propane must be changed to a gas, and air must be added... The gas is then taken into the carburetor where it is mixed with air (and) enters the engine, (where) it is ignited by spark plugs, just as atomized gasoline is fired in a gasoline motor."

CTA orders for propane-powered buses were for years a mainstay of the Twin Coach (later Flxible) Loudonville, Ohio, factory. It shipped 1,400 to Chicago during the era of so-called "old look" buses, the final lot being 100 - 35 ft long x 102" wide vehicles of the 8400-series. They were used on downtown shuttles and neighborhood routes. The last revenue operation of this type occurred on February 15, 1974. One, the former 8439, became work bus BW-49, but only for a short life as the fuel supply was phased out.

Harrison/Halsted - 1-4-73 - A. Peterson

Propane, with an octane rating of 125 (compared to modern unleaded gasoline at 87), was used in a high-compression engine. This resulted in a smoother performing engine than the diesels of the day, with higher power output at low and high speeds. These characteristics made it the most popular vehicle with the public during the critical period of streetcar phaseout.

"..propane boils at –51° F... It has no odor, and must be artificially odorized before it is distributed...so that leaks...may be easily...detected. Unlike gasoline, propane remains in a liquid state only so long as it is kept under pressure. For this reason, it is transported to storage tanks under pressure, transferred (to) bus tanks...and carried in the bus...under pressure.

"Propane is as safe as any volatile substance, and safer than most. For example...in case of fire, flaming gasoline runs and spreads until it is extinguished or burns itself out...Propane, on the contrary can be ignited only when it is a gas and properly proportioned with air. A propane fire does not spread...it will only burn at the source of the fire—in the manner of a blow torch—until the fire is put out or the supply of liquid propane is cut off."

In the spring of 1955 a propane bus was struck by a train and, although cut in two, the safety features operated automatically and there were no injuries or any fire. Notwithstanding, there were accidents later at North Park, Kedzie and 69th garages which dimmed enthusiasm for the propane bus.

Ultimately, the propane technology was not generally adopted by the bus industry although it had some success in the trucking field. At CTA, most of the propane buses were being supplied by Flxible who also happened to be the sole source for coaches with double-stream exit doors favored by CTA staff. Protests by competitor GM Truck & Coach brought about a "test" procurement of buses with single-stream doors, and coincidentally, with diesel engines.

This started the fadeout of the propane era and the beginning of the present diesel era. With newly restrictive air quality standards not far ahead, consideration of alternative fuels is again moving to the fore. One of them is CNG (compressed natural gas), having many qualities similar to propane.

Hermitage/Rogers - 7-28-74 - A. Peterson

FTA-sponsored Superbus project resulted in one M.A.N bus, diverted from Düsseldorf production, being sent to be test driven on several US properties in the summer of 1974, after which it returned to Chicago, was repainted in CTA's special Bicentennial livery and named *Baron von Steuben*. It was then assigned to Route 40-O'Harexpress. During following months many buses and rail cars were similarly repainted and named for American historical personages of the 1776 era. After the Bicentennial, bus colors reverted to the previous two-tone green scheme, while most rail cars were finished in platinum/charcoal with red/white/blue accent stripe.

Northwest Hwy/Foster - 11-28-74 - A. Peterson

High capacity buses

The recurring need to improve the productivity of bus transit, combined with heavy peak traffic demands on a number of routes, led the industry, with the help of the Urban Mass Transportation Administration (now FTA), to study high-capacity technology in 1973. Bus widths had been increased from 96" to 102" and length to 40'; street traffic conditions made it difficult to operate a bigger bus, or so it seemed.

But at the time a number of foreign transit systems were using the double-deck coach or a more recent development, the articulated bus, where two short bodies were joined with a swiveling joint to make a high-capacity single-deck unit.

After some preliminary evaluation, a task force in which CTA was a participant, travelled to Britain, Sweden, Germany and Spain to see what was being done in this field and to talk

To improve customer capacity, relieve street congestion and improve system productivity, some 145 MAN articulated buses were introduced beginning in 1979. They were concentrated on appropriate routes out of North Park, Archer and 77th (later 103rd) garages. Two orders were placed: the first lot of 20 were built in Germany, except for interior finish applied in America by AM General. The follow-on group were built at the MAN's American plant near Charlotte, North Carolina. Bus 7013 got repainted with the trial paint job shown, resembling the livery that had been adopted for rail cars.

95th/Bishop - 10-4-87 - A. Peterson

directly to their counterpart operating officials there about the high-capacity bus experience.

Next, a single German-built (MAN) articulated coach made a nationwide tour in America which demonstrated the practicality of the concept under our street conditions. In Chicago, all doubts as to the ability of the articulated bus to maneuver sharp turns between narrow streets were erased by a couple of runs over tightly kinked Route 157 in the gold coast area (then limited to 35-ft buses) and through the private drive bridge of the Marina Towers, with its sharp horizontal and vertical curves. With the subsequent acquisition of 145 production coaches, Chicago has become one of a growing number of American cities to benefit from this advanced bus technology.

Howard/Hermitage terminal - 3-27-88 - A. Peterson

Air conditioning

Air conditioning is an amenity where CTA experience has been faltering. After a test of only ten vehicles in 1965, the plunge was made with an order of 525 buses in 1972. Additional orders brought the AC fleet to a peak of about 2,000 vehicles by 1982.

Unfortunately, as a siege of funding crises swept the property in the following few years, maintenance of AC hardware became a target for economies. As an economy move, reasoning that Chicago's climate really only *required* bus cooling on relatively few days, management ordered the feature disabled on all but the 145 articulated buses. Unfortunately, when cooling is needed, however often that may be, a Chicago bus is a miserable place to be, so the present board decided that AC shall be restored as replacement buses are acquired. Accordingly, in part of its 1991 bus acquisition air conditioning has been specified. However, at that rate it will require about 12 years to achieve a 100% AC bus fleet.

This 1962-built GMC worked well past its nominal service life—it was a tribute to Foreman Jim Ward's team at Forest Glen garage as well as the South Shops heavy repair group. Replaced finally by Flxible "Metros" in 1991, bus 301 was preserved at Kedzie garage.

Recent bus procurements represent some backdown in omitting air-cooling in hope of stabilizing operating costs, but some improvements were specified, for example, a wheelchair lift at the *front* door of 1991-model Transportation Manufacturing Corporation RTS buses (left). Note also the improved readability of the destination sign as compared to that on the 1986-model M.A.N "Americana" bus (right).

Between 1972 and 1977 1,870 GMC *New Look* ("fishbowl") buses were purchased with federal and state funding assistance. About 500 remained in service by 1992, the remainder replaced by 961 buses received in 1990-1991.
At left, 1972 GMC; at right, 1983 Flyer.

After a bus has served Chicago for more than 20 years it has fulfilled reasonable life expectations, yet in 1991 it seemed that ten of the recently retired 7400-series would have a second career in Zaire. Although this deal failed, at press time 1951-built trolley buses retired in 1973, much rebuilt, are said to be still serving in Guadalajara, Mexico! Also, a number of 1950-built 6000-series rail cars Chicago retired in 1987 are providing backbone service to SEPTA's Norristown High Speed Line.

79th/Western terminal - 11-17-90 - A. Peterson

Handicapped accessibility

In the 1970s and 1980s the added cost and operating difficulty of serving wheelchair passengers caused a general reluctance throughout the transit industry to accept responsibility for serving that small segment of the market, but ultimately the importance of the social objective of doing so resulted in its being mandated.

CTA tested operation by its own forces of a special service, utilizing dedicated, mostly small, lift-equipped, radio-dispatched buses. Such activity proved to be more economically provided by private contractors. It was also not the "main stream" route-oriented service seemingly preferred by the handicapped community, and ultimately mandated by the *Americans with Disabilities Act of 1990*.

South Shops - 6-15-91 - A. Peterson

69th/Dan Ryan station - 10-20-91 - A. Peterson

With the introduction of TMC RTS-series buses in 1991, CTA began its so-called "Main Line" service, in which 54 designated routes are currently operated with these wheelchair lift equipped buses. Initial assignments for the TMCs were at North Park, Kedzie and 77th garages.

56th/Lake Park - 5-10-91 - M. Charnota

The 470 Flxible "Metros" of 1991 incorporate three engine types for comparative performance evaluation. The bulk of the order uses Detroit Diesel 6V92T turbocharged engines (first introduced on rebuilt GMC bus 9532), 20 were equipped with Cummins L-10 power and five with DD 6V91 equipment.

In 1991 scheduled service on regular routes began to use buses equipped with lifts (at the front door) for the better accommodation of disabled passengers. Within a few years, as present vehicles are replaced by new ones, it is evident that all bus routes will be operating vehicles equipped with wheelchair lifts.

The 1991 bus procurement

Of 1991 bus deliveries totalling 961, 491 were Transportation Manufacturing Corporation's model RTS-08, a new design to CTA specifications revised from those developed by GM Truck & Coach, but still featuring a stainless steel modular frame. The remainder were from Flxible Corporation. They were costing up to $202,000 each.

In addition to having a wheelchair lift at the front door, the new buses include driver-operated public address equipment and those built by Flxible have an advanced type of air conditioner expected to be able to stand up better to Chicago service conditions than those of the past. The seating arrangement of the 1991 buses includes a greater proportion of single seats and seats facing the aisle with just a few of the paired transverse seats, formerly so typical. While this is believed to improve passenger circulation in the forward part of a bus, and provides a little more privacy for those who prefer to sit alone, the change has reduced seating capacity up to 30% from past standards, sometimes resulting in standees off-peak where there were none before, as well as overall accommodating fewer people in seats.

A few technical features have been incorporated to making driving more comfortable, but the basic heavy duty diesel power plant with automatic transmission has been retained. Destination signs from now on will probably be of the electronic dot-matrix type, perhaps not quite as legible as the colorful linen or mylar curtains of the past, but easier to modify and cheaper to maintain.

Also, the system-wide standard overall dimensions of 480" length and 102" width are retained, except for 15 of the TMC buses, which are only 96" wide. These are intended to serve Route 16 - Lake Street, where, through a stretch of about four miles, columns supporting the elevated Lake Rapid Transit Route are between the curb and running lanes and a little more clearance was deemed advisable.

In 1987, as part of its new image, CTA introduced for this equipment a red, white and blue livery with grey rocker panels. This has been adopted as the current standard for the whole bus fleet.

◊ ◊ ◊ ◊

55th/Prairie - Circa 1920 - Krambles archive

This view on the architecturally interesting structure at Garfield (55th) station captured three varieties of South Side turn-of-the-century cars, left to right, "Big Sprague" 212, "Westinghouse" 305 and "Small Sprague" 52. Crew requirements in those days included the motorman and conductor plus a guard for each car over two in the train. In 1949, with steel cars and remote-control powered doors, two-man crews were finally achieved. In 1961, one-person operation of single-car trains began on the Evanston line. Today, one person trains provide Skokie Swift service and are anticipated for the route to Midway Airport.

Chapter 7 - 'L' and subway cars

On October 1, 1947 the former Chicago Rapid Transit Company handed over to CTA 1,028 motor cars and 595 trail cars. Except for four articulated cars ordered (with court permission), two of which were just delivered that summer, the newest in this fleet were already in their third decade of service and the oldest dated to 1893.

The original equipment

Over the first half-century of development of rapid transit cars a basic footprint design was adapted from that of railroad coaches of the day, scaled down to fit the smaller clearances of the rights-of-way and tracks of the planned elevated railways. In Chicago, this was a length of 46 ft. (48 ft. on later cars), a width of 8 ft. 8 in. and an overall height of about 12 ft. 6 in.

The body had an enclosed passenger compartment, at each end of which there was an open loading platform with iron wire gates. The interior was fitted with a longitudinal array of seats along each side, relieved in the central part of some cars by two pairs of facing transverse seats on each side of the car. A raised central clerestory was incorporated longitudinally in a so-called "railroad" or "monitor" roof having a number of vertical glazed sash on each side which could be swung outward to scoop in or suck out fresh air as the car moved.

Wood was the principal structural material, reinforced with iron truss rods to hold the floor frame level. The cars rode on iron framed, four-wheeled trucks.

Later wooden cars

Over the years car designs were improved to include enhanced comfort and safety features such as enclosed (instead of open) platforms, sliding (instead of folding) doors, electric (instead of coal-fired hot water) heaters, and reversible (instead of fixed) transverse seats.[1] The application of intake and exhaust ventilators made possible the adoption of the plain arched roof, eliminating the leaky ventilator sash. Steel-reinforced underframes had been included in cars built after about 1904, although the basic

---

[1] The pros and cons of fixed vs. reversible and longitudinal vs. transverse seating continue to affect transit car design. No solution seems to satisfy all goals of safety, style and passenger acceptance.

Cincinnati Car Co. - 1914 - Krambles archive

In 1914, the 4000-class marked the change to steel cars. Funds were short, so the first 66 were built as trailers, but were designed to be fitted with motors in the future (only one, 4005, was). Follow-on orders for 389 motor cars eventually swelled this classic group. Interestingly, during early CTA years, to improve performance a number of *wooden* trailers were motorized, since it was more urgent to get rid of trailers on lines other than the State Street subway where the 4000s were then concentrated.

Train consists in pre-CTA and pre-subway days were a mix of old wooden-bodied motor and trail cars with one or two steel cars. In peak hours, trains on the west side would often be 2M + 3T makeup. Here, westbound at the curved Sacramento, Lake Street station with its characteristic "pagoda," 1924-vintage car was on the point. Sacramento was one of ten stations closed in 1948, when the A & B plan started.

Cars 4251-4455 had many refinements over earlier 4000s. More windows could be raised, there were circulating fans on the bulkheads, light fixtures had opal shades, there were emergency battery lights, canvas-covered wood roofs, and trolley poles as well as third rail shoes. There were more reversible seats and they were covered with plush.

Sacramento/Lake - 10-24-42 - J. J. Buckley

construction material was wood. Over the years the underframes of a number of all-wood cars were similarly rebuilt.

Steam locomotives

The mechanical and electrical features of cars also changed gradually over the years. The first 'L' trains on the South Side route consisted of trailers, pulled by tiny steam locomotives of a rather sophisticated Vauclain 4-cylinder cross-compound design built by the Baldwin Locomotive Works. The first Lake Street route trains also used compound steam locomotives, but of a two-cylinder design built by the Rhode Island Locomotive Works. Both lines used vacuum braking, steam heat and Pintsch gas car lighting in those days.

Early electric motor cars

Rapid development of railway electrification technology obsoleted steam power by 1895, when the first Metropolitan route began operation. Their first trains consisted of a motor coach with two motors in one truck, pulling one, two or even three trailer coaches. Straight air braking was fitted. The Lake route changed from steam to a type of electrification similar to the "Met" in 1896, rebuilding some of their trailers into locomotive cars at that time.

Wilson Shop - 1922 - Krambles archive

Grove/Evanston - 11-6-73 - A. Peterson

Conversion of the Evanston branch from overhead trolley to third rail was coordinated with replacement of half-century-old 4000-series cars. New aluminum-clad contact rail allowed later removal of unsightly overhead cable and most of the wooden line poles. One day after this photo, run 511 leaving Linden terminal, Wilmette, at 0856^2, made the last run of 4000s in scheduled service and also, on its return, was the first to operate by third rail north of Howard. Because of the energy crisis at the time, 74 - 4000s were mothballed for a few months, but were never needed for passenger service. Others had already replaced wood-bodied cars as work train motive power.

Sedgwick/Ravenswood station - 3-2-80 - A. Peterson

Cars 4271-4272 in mid-1972 became the last of their series to be overhauled. Chosen to become a historic train, some additional work, including the reminiscent paint job and an adaptation of the cab signal system, was done in 1974. No other 4000s were ever equipped for cab signal operation and, as a safety precaution, all but 4271-4272 were embargoed from use as driving cars even in work train service effective 0900 July 17, 1977. MCERA John F. Humiston, engineer, with a team of ingenious technicians at Skokie shop, developed a practical means of adapting all-electric 6000-class cars, which have ATC-cab signals and digital train line controls, for use as locomotives to work with analog-controlled air brakes on trailer work cars.

The South Side line had been a pioneer in utilizing the 4-cylinder compound steam engine, but it innovated much more importantly when it switched to electric traction in 1897-1898 by having Frank Julian Sprague's just invented multiple-unit control, motors and the appropriate train wiring, together with automatic compressed air brakes, electric lighting and heating retrofitted onto its new coach fleet.[2]

Of course, there were many other technical changes over the years before CTA: better couplers, braking, heating, lighting and door control are just a few of the more noticeable ones.

Steel cars come to the fleet

The best of the standard heavy rapid transit car were Chicago's 4000-class steel cars, designed in anticipation of the unified operation of the four underlying companies that began in October 1913. The first version, cars 4001-4250, appeared in 1914–1915. In later years this type affectionately became known as the "baldy" because of its largely unadorned plain arch roof. In all, there were 455 of the two types. They incorporated lessons learned from the first quarter century of Chicago rapid transit experience in safe, dependable operation and economical maintenance with good collision absorption and fire resisting qualities.

Yes, there was a stark plainness of the "baldy" design, resembling the coldly efficient designs then cropping up in the Boston, New York and Philadelphia even to the provision of a side center door, which as it turned out was not used. About half of this group had the "bowling alley" (longitudinal) layout of rattan covered seats. The 1915 cars seated 52 passengers, 32 of them on reversible transverse seats, unfortunately not exactly lined up with the windows. Actually, the fold-up longitudinal seats along the unused center doors had hardly any view.

Chicago's most attractive steel car

The second, more advanced model, represented by cars 4251–4455, came into service in 1922–1924. Its luxurious interior and green plush seats earned it the sobriquet of "plushy".

There were many crowd-pleasing changes incorporated in the "plushies". As with the previous steel cars, there were seats for 52, (36 on reversible seats accurately located at the windows). All were covered with green frieze plush. Other refinements included circulating fans on the bulkheads, more windows that could be opened, and opal shades on the ceiling lights. "Plushies" also had a few battery-powered ceiling and vestibule lights that came on automatically when the regular 600-volt lighting was interrupted, as when crossing from track to track or running through a street crossing.

The leather "straphanger" belts overhead were gone and in their places were neatly spring-aligned, porcelain enamel hand-holds. The steel bulkhead panels were finished with expertly simulated mahogany-wood-grained paint. The floors were of maple instead of the cold, concrete-like red composition surface of the older steel cars.

To top it off (no pun intended), the roofs were of wood, covered with canvas, with an increased number of mechanical ventilators, and carried trolley poles. These, together with various electrical modifications, made this final group of heavy steel cars suitable for operation anywhere on the Chicago system. Although they represented only about one-eighth of CRT's fleet, at least a few were assigned to each division. For twenty years, they were arguably the finest rapid transit cars in America.

Pressure builds to get new cars

CRT had been under great pressure to replace nearly 1,200 wooden cars which, in addition to presenting a special hazard on the largely unsignalized lines, represented the teething days of electric railroading and were technically obsolete. For years the issue had arisen after every collision. Before Chicago's State Street subway, its first, was built it was agreed between the city and CRT that only all-steel cars would be operated in it.[3] There were no funds with which to buy new cars, so when the time came, all 455 of the 4000-class steels had to be rounded up for that busy line.

Over the years, studies were made to explore how system productivity might be improved by adopting a larger vehicle, like those on the current extensions in Boston, Philadelphia or New York. The CTA ordinance enumerated the related necessary work as a modernization task.[4]

However, throughout the main lines and yards of the Chicago system there are sharp curves and clearances that were adopted to solve right-of-way and access problems. Modifying

---

[2] MU control permits any number of motorized coaches coupled into a train to be driven simultaneously from any operating cab. Making a long, high performance train feasible without requiring a heavy locomotive, this concept accelerated electric railway development around the world.

[3] The CTA Ordinance of 1945 affirms this; quoting from Section 9, Paragraph E: *"All passenger cars...operated in...subways shall be of steel or other comparable material, and of approved construction for the safety, comfort and convenience of the patrons of the Transit System."*

[4] *cf* CTA Ordinance, Exhibit B, Item 1 - Rapid Transit System if Acquired by Grantee.

203—Train Platform, Chicago's Initial Subways

1943 - Curt Teich & Co.

Racine/Met (Congress St.) - 5-24-49 - Krambles archive

the many platforms already built to meet the narrow (8'-8") floor line would have been an overwhelming cost problem. For these and other reasons, it never proved economically feasible to use a substantially longer and/or wider car. That is, not until the Chicago North Shore & Milwaukee Railroad's *Electroliners,* built in 1941, proved successful in running over the 'L' with their concave-convex side section providing a width of 9'-4" at the belt rail while staying 8'-8" wide at the floor. Those unique trains also demonstrated another technique for increasing the capacity of a rail vehicle—articulation.

As Chicago's first subways neared completion, commitment in principle was made to the articulated PCC concept that Clark Equipment Company had developed for Brooklyn's "Bluebirds". This postcard, issued on opening the State Street route in 1943, was an artist's concept of one. Later, rapid transit in Cleveland and Boston also used PCC technology. Chicago surpassed all others, using 774 PCCs on rapid transit. At this writing, Philadelphia's Norristown route was the only other rapid transit application of PCC cars, theirs coming from Chicago.

In 1947-48 Pullman and St. Louis each built two artics to Chicago clearances. Here they were on a test run near Throop Street shops. where three levels were interconnected with an elevator–the only place where an inboard truck could be changed without disassembling the articulation; there were no drop pits in those days. This led to preference for the married pair concept when production runs of cars followed. The 5000s languished for years on Evanston and later Ravenswood. Their productivity potential belatedly showed when assigned to Skokie Swift.

Further, the application in new rapid transit cars of the PCC technology that had proven successful in streetcar service was now recognized. The souvenir brochure for the October 17, 1943 opening of the State Street subway included a photo of the sample articulated PCC-equipped subway car that the Clark Equipment Company had produced for the Brooklyn Manhattan subway, with a statement implying that similar new cars would have to be acquired before route #2 (the Dearborn subway) could begin service.

Aluminum articulated car concept is tested

Ready to innovate, CTA agreed with the Department of Subways and Superhighways[4] to try an aluminum-sheathed body with a steel-reinforced frame car utilizing all-electric PCC technology. It would be similar to the Brooklyn sample, but adapted for the smaller Chicago loading gage. DSS initiated an order for 30 cars with St. Louis in 1944, but this order was cancelled when financing could not be completed.

Ultimately, with court concurrence, orders for four were executed, two each from Pullman-Standard and St. Louis Car, with common principal exterior and interior dimensions and construction concepts. They were paid for with CTA funds acquired from the CSL takeover. Engineers of CRT, CSL and DSS worked together to develop the specifications.

Car 5001, the first of the Chicago articulateds, was delivered to CRT in August 1947. Cars 5001–5002 were built by Pullman-Standard and were equipped with Westinghouse propulsion package and Clark trucks. Cars 5003–5004 were built by St. Louis Car with General Electric propulsion and St. Louis trucks. By the time delivery was completed from St. Louis CRT and CSL had been absorbed into CTA. After brief trials in revenue service on all the principal rapid transit lines, the 5000s were settled into a regular assignment in Evanston operation.[5]

But the married pair standard is adopted

The matter rested there only briefly as the new owner took over with a running start in tackling the modernization problem. The same engineering team that had breathed life into the articulated car concept were already at work for CTA. They prepared for a production order of new cars, and, while they liked their handiwork on the 5000s, they decided it could be improved by some changes, most apparent of which was the adoption of the married-pair body configuration, with two single-end cars permanently coupled back to back.

---

[4] At that time DSS was the City of Chicago department responsible for design and construction of the State and Dearborn subway routes.

[5] Later, the 5000s spent several years on Ravenswood, but they really hit their stride 1965-1985 as Skokie Swift cars 51–54.

Better air comfort was sought (in vain) with special ceiling fans on car 6047 and evaporative cooling on 6669, shown here. Air conditioning came with the 2000-series. Assuming that future cars would have fixed side windows and air conditioning, close side clearances were agreed in engineering the extension to Jefferson Park, but initially 6000s would have to be used. To prevent arm-out-of-window accidents, side windows could be sealed closed if air conditioning were installed. Car 6711 was so fitted, but the cost and time-to-do proved prohibitive, so window bars were installed instead.

Armitage/Ravenswood station - 8-26-59 - R. G. Benedict

Cars 6001-6200 were delivered with the conductor's working location outside between the married pair cars. Operating experience with the first six cars showed a need to improve the conductor's view into adjacent cars, "standee" windows were provided at the #2 end of each car. Ultimately, the conductor's position on this group of cars was relocated inside the "A" car of each pair.

Both: Logan Square yard - 9-16-50 - G. Krambles

By September 1948 (only a month after the delivery of the last of the 5000-series) the project had progressed to an order for 65 two-car sets of a new specification, the 6000-series cars. This design went through several iterations during the period 1950–1960 as a total of 770 cars (including cars 1-50) were produced, all by the St. Louis Car Company.

The first 200 of the 6000s were built from totally new materials designed for the purpose. The remaining 570 were built using components recycled from 570 PCC streetcars built in 1946–1948 and still in service on the surface system at the time.

Hardly had these PCCs been delivered when it was decided to phase out streetcars. Rapid growth of the automobile population was causing rapid loss of riders (especially from the surface lines) and at the same time making streetcar operation in the middle of narrow streets slow and accident prone. Operating cost could be cut by driver-only bus operation. Money could be saved by avoiding renewal of deteriorated tracks and other street railway plant. Coupling this with the need to upgrade the rapid transit fleet yielded the concept of remanufacturing streetcars into rapid transit cars.

Completely new body shells were built with strong underframes and couplers, and certain other components were modified to make them suitable for train operation. After reconditioning, it was possible to utilize not only the major components such as motors, propulsion, batteries, trucks and motor-generators, but also furnishings such as the window sash and mechanisms, seats, lights and door mechanisms. Over all, much value was saved by the process.

Characteristics of series 6001–6200 are:

Length of unit over coupler pulling faces	96'-6"
Width at floor	8'-8"
Width at windows	9'-4"
Height over roof	11'-10"
Trucks	Clark B-2
Truck centers	33'-8"
Truck wheelbase	72"
Wheel diameter	26"
Coupler: #1 end / #2 end	Form 5 [6] / tubular
Seats: A car / B car	47 / 51
Weight per car, without passengers	41,700 lb
Motors per car	4 W 1432 @ 55 hp
Propulsion control	PCC accelerator
Free speed	50 mph
Lot cost (first 200 cars)	$7.85 million

Some details of later cars 6201-6720 varied, but all 774 cars of the 5000, 6000, and 1–50 series could be intercoupled into trains of appropriate length for the required service.

Cars for one-person single-car operation

The fifty cars of series 1–50 were equipped with cabs and form 5 couplers at both ends, for double-end operation singly or, in trains of one or more cars, together with the 6000s. Side door locations and interior details were laid out to permit driver-only operation.

Some cars were fitted to accommodate overhead trolley equipment in anticipation of future use on the Evanston branch and, as it turned out later, on Skokie Swift.

Closing the decade of acquisition of this type of equipment was the 1960 delivery of cars 1–4. Fitted with developmental high-speed propulsion packages, they were planned as a step in developing specifications for the next generation of equipment. Spectacular in performance (up to 70 mph) as compared to their siblings, they reached their career pinnacle when assigned to Skokie Swift from 1964 to 1974. With the extra equipment required, the weight of cars of the 1–50 group ranged between 23 and 26 tons.

Meanwhile, back on the railroad

Even as the 6000-series was under design, engineers were working out ways to improve the performance of the old car equipment received from CRT. Financial resources for fleet replacement with *new* vehicles had yet to be developed, so the old equipment would have to serve for an indeterminate future.

Identified for early action were:

√ - *Elimination of wood-bodied trail cars to improve performance and safety,*

√ - *Installation of multiple-unit door control on 4000-series cars to improve safety and reduce crew cost, and*

√ - *Reduce the manpower needed for incidental duties in train operation.*

Changes in operation plans, for example, the closing of stations with light usage and the introduction of 'A' and 'B' service, reduced the maximum number of cars required to fill schedules. Wood trailers were the first to be retired. A few of the best trailers, those with air engine doors, were motorized using truck, motor and control components that had been in storage for more than twenty years.

The purchase and installation of hardware to make MUDC possible, i.e., to enable control of all side doors of 4000-series trains from any one car, did for door operation what Frank Sprague's MU invention had done for control of propulsion 50 years sooner.

To maximize capacity and minimize weight in the 6000-series then under order, a concept had been adopted of marrying cars into *permanently* coupled pairs. The idea was applied in modified form in retrofitting the 4000s: they were married into *semi-permanently coupled* pairs. The completed installation permitted a train of any length to be crewed with just two men. Five had been required to handle an eight-car train. The total cost of the work was returned by savings after about nine months, a superb economy!

A number of incidental changes were made to reduce running and standing time and to reduce the manpower needed for incidental duties in train operation.

Control of the 4000-series, previously energized from the 600-volt system, was now changed to battery feed, reducing power loss when snow, frost or ice coat the third rail.[7] Tying together of the overhead trolley feed between cars of a married pair assigned on the Evanston route made it possible to operate with only one trolley on the wire to power both—a big advantage in changing between third rail and trolley, then required on this line.

The old hand-operated destination signs in cars were changed to indicate both terminals so they would not have to be manually reset for

---

[6] Form 5 is a knuckle coupler like the familiar MCB used on railroads, but smaller.

[7] In hope of improving braking rates, increasing brake pipe pressure from 70 to 80 psi was tried on the 4000-series cars, but the test was abandoned after a few months due to slow and uneven release on the 30-35 year old cars.

Willow/Bissell - 8-9-91 - A. Peterson

Rosemont/O'Hare line yard - 2-11-91 - A. Peterson

The 180 1964-built 2000-series marked a sharp break from the 6000-series PCC family in technology of trucks, motors, control, carbody, lighting, air comfort and even styling. With so many new features, they've challenged engineering and maintenance forces over the years.

During 1990-1992 the 144 remaining Budd cars built in 1969 were rehabilitated by New York Rail Car Corporation, with up to 24 cars off the property at a time. The motors of ten were replaced with new ones of an Alsthom design which had been successfully pre-tested on cars 2250 and 2566.

each trip and the number per car was reduced from four to two. Offsetting this, massive additional amounts of fixed graphics were added on station platforms, and loudspeakers were provided on station platforms and in all new cars. Platforms were lengthened so that all side doors could be used for passenger interchange at every station.[8]

High-performance cars from Pullman

By 1960 all usable PCC streetcars had been remanufactured into subway-'L' vehicles; studies and component tests had been completed. The several carbuilders then serving the American market had been consulted and had offered their suggestions. The resulting 2000-series design specification, calling out for the new generation a high-performance car, was put out to bid. It called for alternate bids, one to include

---

[8] Except, of course, the right-front door, which was part of the driving cab.

air-conditioning. This was hardly more than a tongue-in-cheek option, as previous experience indicated it might add unsupportably to the cost, but CTA was happily surprised to learn it added only about 7% to the price. Thus came air conditioning to CTA rail.

Pullman Car & Manufacturing Company's winning offer was for 180 cars (90 units), series 2000-2180. These introduced other innovative features such as sculptured fiberglass ends, large picture windows, AC-powered fluorescent lighting, remote-controlled destination signs, aluminum and end-grain balsa wood floors, solid state battery-charging convertors, and a new truck design developed cooperatively by CTA engineer Charles E. Keevil and the LFM (later Rockwell) Company.

Since the propulsion package was significantly more power than that of the 6000s, a different arrangement of electrical coupler was chosen to become a new standard for this and future orders. This permits mechanical, but not electrical, coupling with the older equipment, a handy capability for emergency movement of a disabled train.

The interior was unpainted, utilizing instead anodized aluminum, fiberglass and melamine finishes. Styling details were advised by consultant Peter Muller-Munk.

Principal characteristics are as follows:

Length of unit over coupler pulling faces	96'-6"
Width at floor	8'-8"
Width at windows	9'-4"
Height over roof	12'-0"
Trucks	CTA-1
Truck centers	33'-8"
Truck wheelbase	78"
Coupler: #1 end / #2 end	Form 5 / tubular
Wheel diameter	28"
Seats: A car/ B car	47/ 51
Weight per car without passengers	47,400 lb
Motors per car	4 GE1250K1 @ 100 hp
Free speed	55 mph[9]

Shortly after delivery of the 2000s, installation of continuous automatic train control with cab signal was begun over the almost all main line trackage then unprotected with signals. The 2000s were the first type of car to which the related carborne equipment was added.

Stainless steel cars come next

Operating and maintenance experience with the 2000s led to incremental change in specifications for the next acquisition, which was for 150 cars (75 units), series 2201–2350.

The contract was won by the Budd Company and was built in its Red Lion (Philadelphia) plant. Skidmore Owings & Merrill, design architects for the Dan Ryan–Kennedy extensions, served as styling detail advisor. The cars were delivered in 1969.

Probably the most significant change that appeared with the 2200s was the use of unpainted corrugated stainless steel (instead of smooth aluminum) for side sheathing and roof, a Budd-pioneered concept in rail car construction which was competitive with aluminum in weight saving as well as being more fire-resistant. Responding to the styling preferences of the architectural consultant, the side cross-section was redesigned with flat vertical surfaces and a sharp taper near the floor line instead of the convex side configuration of previous (and subsequent) orders.

Air conditioning was improved by using a system of ducts to distribute air more uniformly than had been the case with the ceiling-suspended package of the 2000s.

Another innovation was the application of the Budd Pioneer III truck, originally with hollow axles and aluminum centered wheels. Other changes were mainly cosmetic in nature or were incremental technological improvements.[10]

The procurement of the 2200-series was unique in being arranged by the City of Chicago as part of the extensions of rapid transit in the Dan Ryan Expressway to 95th Street and in the Kennedy Expressway to Jefferson Park. They were also the first Chicago cars to be funded in part through a grant from the U.S. Department of Transportation, Urban Mass Transit Administration, today's Federal Transit Administration.

Aircraft manufacturer builds the next cars

Time was fast approaching for the retirement of the last of the 4000-series steel cars inherited from CRT, then already well into their fifth decade of passenger service. Minimum fleet requirements for peak service were barely met by the combination 2000s, 2200s, 1–50s and 6000s, but the latter were starting to show their age too. As quickly as funding could be arranged utilizing grant assistance programs of FTA and the State of Illinois, another specification for cars "hit the street". It was for 100 (50 pair) of the 2400-series cars with an option, ultimately exercised, for a further 100. The order was won by Boeing-Vertol, who would build it in its helicopter plant near Philadelphia.

---

[9] The power package is capable of 70-mph, but is limited by the logic of the car controls and the external signal system. As built, a speedometer was applied in the cab of cars with numbers ending in 1 or 5. Today every cab includes a speedometer integrated with the automatic train control cab signal display unit.

[10] One change that didn't pan out: as built, the door, seat and back wall of the motorman's cab could be folded out of the way to make an open seat available for passengers when the cab was not needed for driving; this is no longer used.

Mannheim/I-190 - 9-17-84 - A. Peterson

A solid string of matched cars, as shown eastbound from O'Hare International Airport, is pleasing to the eye. However, a car from any series can be operated on any line and in the same train with any other car from the same family. Cars of different families (i.e., a 2000 or a 6000) can be coupled mechanically, but not electrically.

FTA-sponsored State-of-the-Art (SOAC) cars to demonstrate best 1970s ideas for potentially "standard" subway cars ran for a couple of weeks on several properties. Based on the then-current New York "B" division car, they were too long and wide to clear most of the Chicago system, but a test was set up on Skokie Swift. To handle both SOAC and CTA cars, motorized flaps were provided on one platform at Howard and both platforms at Dempster.
A single-track was operated between Howard and this crossover.

Ridge/Skokie Swift - 1-19-75 - A. Peterson

It was a time when aircraft builders were seeking new markets for their skills; the Rohr people had landed major car contracts with the Bay Area Rapid Transit District and with the Washington Metropolitan Area Transit Authority. Not to be outdone, Boeing-Vertol had taken on an even more challenging job to supply "new standard" articulated light rail vehicles in a joint procurement for service in Boston and San Francisco.[11]

Most of the changes from the 2200-series were again incremental, with principal dimensions essentially unchanged. Styling consultant for this contract was Sundberg-Ferar, who created a sculptured mask for the #1 end. It was

---

[11] Ultimately, 175 cars were ordered for Boston and 100 for San Francisco.

manufactured of molded fiberglass, a novel technology first used on the Pullman 2000s. Sides were of smooth stainless steel accented by two horizontal ribs. Roofs utilized stainless steel, corrugated for longitudinal stiffness. Applying its aircraft builder's expertise in worldwide procurement, Boeing subcontracted the complete body shells to Sorefame, once a Portuguese firm and Budd licensee, and the trucks to Wegmann, a German firm.

Chicago's long application of the fast-working blinker side door was superseded by a two-part sliding door providing the wider opening needed to accommodate a wheelchair, in the fast rising recognition of special requirements of the handicapped. The motorman's cab became dedicated crew space following an operating decision to no longer make unoccupied cabs accessible to passengers for security reasons.

The basic all-electric propulsion package was the same as had been supplied with the 2200s. Seeking to graduate from the electro-mechanical camshaft (SCM) control, it was specified that 10 cars should have chopper control, a solid state system expected to reduce maintenance and to save energy through regeneration during braking. However, the contract was conditioned upon satisfactory completion of acceptance testing of prototypes. This revealed unanticipated electro-magnetic interference, threatening the integrity of signal track circuits. In the end, all cars were completed with SCM equipment. Adoption of more advanced propulsion technology at CTA is a probability that awaits future development.

The largest order of CTA cars ever

With access to capital resources better assured by the addition to FTA and I-DOT of the Regional Transportation Authority with some local sales tax entitlement, the catch-up process in replacing life-expired rolling stock continued with a December 1978 order. Due to incremental funding needs, initially 300 cars to be delivered over five years were specified, but this was increased to a total of 600 (300 married pairs).

The prospect this huge order created for years of continuing work attracted bidders. It seemed likely that Boeing-Vertol, which had already mastered the CTA's basic requirements with its just-completed 2400-series work, would offer the best bid, but surprisingly to the contrary, the most favorable bid was that of the Budd Company, later to become Transit America. The work was done at the Red Lion plant.

Budd built the car shells themselves as well as doing the assembly, but used some European-made components such as the Wegmann trucks and Krupp motor-alternators. Again there were incremental changes based on experience with the 530 high-performance cars of earlier orders and available technological improvements in component parts. For example, interior floor space was slightly increased without changing overall dimensions by redesign of the car end contours. A dedicated space and a folding seat for wheelchair tie-down was provided, although at the cost of one pair of seats in each unit.

Principal characteristics are as follows:[12]

Length of unit over coupler pulling faces	96'-6"
Width at floor	8'-8"
Width at windows	9'-4"
Height over roof	12'-0"
Trucks	Wegmann
Truck centers	33'-8"
Truck wheelbase	78"
Wheel diameter	28"
Coupler: #1 end / #2 end	Form 5 / tubular
Seats: A car/ B car	43 / 49
Weight per car without passengers	54,300 lb
Motors per car	4GE1262A1 @ 110 hp
Propulsion control	GE type SCM
Free speed (governed)	55 mph

The cost of this lot of 600 cars was $323.7 million. With the completion of the CTA order, Transit America retired from carbuilding.

Remanufacturing and upgrading

In 1985-1986 45 of the series 5-50 cars (assigned to Evanston-Wilmette and Skokie Swift service) were rehabilitated and upgraded for an anticipated additional 10-year life in a program performed by the Morrison-Knudsen Company, a respected American construction engineering firm, diesel locomotive builder and recent entrant into the transit carbuilding field. The average cost per car was $188,000. Ten are paired off to become 61-65 series two-body units replacing 5001-5004 which were then retired.

A somewhat similar program has been completed on the remaining 72 pairs of 2200-series cars by the New York Rail Car Corporation. The average cost per car will be about $292,000. Of number of improvements being incorporated in the cars the most visible is the introduction of hopper type windows to afford back up ventilation in case of air conditioning fault.

Another order, another builder

The new Southwest rapid transit route extending to Midway Airport, scheduled to go into service in 1993, will require about 100 cars, a system fleet increase of 80-90 cars. This new need, plus approaching life expiration of the few remaining 6000- and 5–50-series as well as a portion of the 2000s, resulted in the January 1990 order for 256 cars (128 units), the 3200-series. Morrison-Knudsen won this contract, having an expected value of $207.7 million, or about $811,000 per car.

---

[12] Most of the data in the table for the 2600s apply equally to the 2400- and to the 3200-series cars.

Typically over the years each new Chicago rail car order included incremental, rather than radical, changes from the previous lot. The current 3200-series, for example, have high-tech microprocessor control logic and diagnostics. The interiors of these cars include more open floor space with less seats and their cabs and door controls are arranged in anticipation of one-person train crewing.

Both photos: Florence/Skokie Swift line - 9-27-92 - A. Peterson

Characteristically fo CTA, an incrementally modified vehicle is specified, fitting established footprint and clearnce limits. Although camshaft control is retained, there are a number of changes to the technical details of component hardware. Examples include incorporating solid state logic and microprocessors for propulsion control, fault diagnosis and automatic train control cab signalling. Slip-slide control logic is added, but sanding is omitted. Duewag trucks built to the Wegmann design are used.

Stainless steel is employed for frame and sheathing of the car bodies, which are produced in the Mafersa (a Budd licensee) plant in São Paolo, Brazil. Fluted side sheathing, unadorned with paint or striping, is used to discourage graffiti marking.

Passengers will notice the two-one (as compared to long standard two-two) transverse seating. This will test market reaction to the tradeoff between the number (and privacy) of seats and the amount of standing space. In surrender to the foibles of air conditioning maintenance, hinged hopper-type upper sash are fitted to side windows, as with the 2200-series rehabs. In anticipation of driver-only operation, full width cabs, similar to those in the 60-series, are provided.

The 3200-series can be interworked in trains with any of the several 2000-3200 series. Among these types, if past practice is followed, the groups will be reassigned around the system from time to time to optimize operations.

At press time, only Skokie Swift is served by non-air-conditioned, CTA first-generation cars. Looking ahead, the procurement of further high performance rolling stock to replace these and the aging 2000s, now rapidly approaching the end of their life cycle, will undoubtedly be high on the capital renewal priority list.

◊ ◊ ◊ ◊

52nd/Prairie crossover - Circa 1914 - Krambles archive

For years after 1947, on-sight operation on the largely-unsignalized CRT network was dependent on alert train crews and wayside employees. Illustrated in this view are two diamond-with-bulls-eye spacing boards, only used in Chicago, to guide motormen in maintaining safe following distance when closing in on a train ahead.
(See also photo on page 30.)

The rigid bow on the roof of the Sprague car was used to collect power from a slack trolley wire used in the yards at 63rd/Calumet, a system converted to third rail to accommodate the mix of cars that followed the 1913 through routing between North and South Side lines.

Chapter 8 - Signalling on the rapid transit

One of the major shortcomings of the rapid transit system inherited by CTA in 1947 was the lack of signal protection to prevent rear end collision where at times of heavy traffic trains might follow one another at intervals of only a few seconds. More than 200,000 trains per year were operating over some 200 track miles most of which were "dark" (without signals for rear end protection).

Over much of the system trains had to operate *on sight,* depending on their driver's comprehension of, and compliance with, established rules for safe following appropriate to the speed of the train. This is the way in which street traffic, including streetcars, has always moved. In those days trains did not have speedometers, but there were fixed marker signs called "spacing boards" in the footwalk between tracks to show motormen the minimum distance to maintain when following a moving train.

Interlocking

Junctions and crossings were fitted with *interlocking*, that is, a group of switches, track trips and signals, with locks so arranged as to permit train movements without conflict.

The oldest interlockings were entirely mechanical, depending on ingenious arrangements of bars and slides between the operating levers to prevent setting up a lineup of switches and signals for conflicting train movements. The mechanical system also constrained the possibility of throwing a switch under a train.

The signal was a semaphore, with a bright red arm facing approaching trains. Its normal horizontal position indicated *stop;* at an angle it indicated *proceed.* The trip was a **T**-shaped device in the center of a track adjacent to a signal. Unless the interlocking was aligned to permit safe passage across the related switch (or crossing), the trip was raised where it would engage a valve mounted under the head car of each train to vent the brake pipe, causing an emergency stop. Only with the trip lowered could the associated signal be cleared.[1]

---

[1] A center trip was of cast iron, painted red. If passed *up* it would be smashed by the first motor casing, clear evidence of rule violation.

Unattended operation, fixed routing

Unattended interlocking is a common cost-saving measure applied to both transit and main line railroads. Chicago's rapid transit experience with unmanned interlocking towers goes back to early mechanical plants at locations like Rockwell/Lake and Barry/Sheffield, where crossover moves were required only in weekday rush hours and the switches and signals remained undisturbed (usually aligned for straight-through moves) at all other times.

Multiple route automation

The first application at CTA of multiple-route automatic operation began in February 1953 at the south terminal of the Logan Square service, LaSalle/Congress.

This installation consisted of a diamond crossover in approach to a two-track stub terminal. Inbound routing was based on which track (if either) was unoccupied, with preference to the crossover move if neither track was occupied. Outbound routing used "first-in, first-out" logic, triggered by an automatic train dispatching program in the control center about one-half minute before the actual scheduled leaving time. This preempted control of the plant to avoid delay to a departing train by the potential coincidental arrival of another train and gave time to complete any necessary switch movements. Finally, a timing device cleared the departure signal and triggered the starting lights and bell at precise schedule time. Controls were retained for conventional local operation, as in emergencies or in rush periods when special routings might be required.[2]

Another innovative step occurred in August 1954, when fully automatic operation was instituted at a new interlocking plant installed the previous April at Lake/Paulina.[3] The layout was a flat junction of two double-track elevated routes. The novel feature was the method of sorting westbound trains approaching the diverging switches using a new electronic system. This required a removable coil to be inserted in a bracket on the front right corner post of cars of the diverging (Douglas) route which interacted with a wayside-mounted coil to request the diverging route.[4] Lake route trains continued without the coils and when one of these approached the interlocking plant westbound, the circuit logic,

---

[2]) Circuitry for such unattended operation has been provided for all terminal interlocking plants installed since, although at this writing not all are being so used.

[3]) This junction was needed during construction of the new alignment in the Congress Expressway from 1954 to 1958 as part of a reroute of Douglas trains from the Loop via Lake Street and thence south via former Logan Square 'L' tracks to rejoin the retained Douglas route at Harrison near Paulina Street.

[4]) The system, a proprietary product, called *Identra*, had seen some use overseas for train routing, but this was the first North American application. The 18" diameter Identra coil was a passive tuned coil attached to a marker light bracket. Its appearance earned it the nickname "toilet seat."

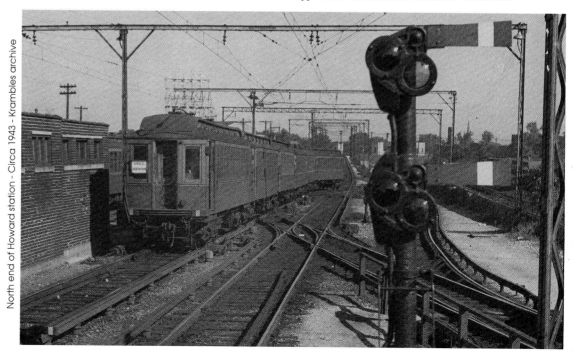

Signalling at interlocking plants when CTA took over was predominantly of obsolete design, like the equipment shown here. Electro-pneumatic switch and signal movements predominated. Enforcement of restrictive indications was by means of track trips located in the center of tracks acting against a brake pipe vent cock on the cars under the coupler. Generally, the trips were not far enough in advance of fouling points to provide worst case protection.

North end of Howard station - Circa 1943 - Krambles archive

Westbound Lake train, #18 Tower, Lake/Wells - 3-23-92 - A. Peterson

In early CTA years a successful concentrated effort was to simplify train routings so that the number of attended towers could be minimized. With follow-on improvement of signal and interlocking hardware plus CTA's unique line supervision system, most terminals were fitted with automatic interlockings that provided first-in first-out train departures according to exact schedule and yielded substantial personnel savings through towermen not needed.
One of the least painful economies, changing train lengths from four cars on 5-minute headways to six cars at 7^2-minutes, saving one two-person crew trip for every 12 car trips. This change of late 1990 also narrowed the period of A-B express service as compared to earlier years. From a marketing point of view, however, the downside is that it costs riders added average trip time.

sensing the absence of the coil, provided the Lake routing. Eastbound (converging) routing was on the "first-come, first-served" basis, that is, the first train to approach would get the route. Wayside route selector pushbuttons were provided for overriding the automatic route selection, as for a missing coil or a car equipment transfer.

Sorting by sequence

When the interlocking equipment at 59/Calumet was modernized in April 1965 the concept adopted for sorting trains utilized the fact that although, as at Lake/Paulina this was the flat junction of two double track lines, this time they were lines with *equal* service, alternate southbound trains going to the 'A' (Englewood) and 'B' (Jackson Park) branches. The interlocking circuitry was designed to remember which way the last train went and to send the next one on the other route. There is remote control in the attended tower at nearby 61st Street yard to deal with irregular moves.

Sorting by transponder

Another approach in automatic train routing began in September 1969 with the Dan Ryan service. This new route joined the existing 'L' Loop at #12 tower, Van Buren/Wabash, again basically two double-track lines joining in a flat junction, but (later) with a track for irregular special moves between the converging lines. Trains of the Lake-Dan Ryan (West-South), Ravenswood and Evanston services pass through this junction on one track or the other, but they are on independent headways and their service periods are unequal.

Convergence is handled again by "first-come-first-served" logic. However, to provide automatic sorting of diverging trains, transponders are installed on cars of the Evanston route and their input sorts the trains as needed.[5] Wayside route selector pushbuttons provide override if and when needed.

Automatic Block Signals

Over the years a few installations of automatic block signals had been made under predecessor companies to provide some rear end protection around curves and other locations where unusual hazards were perceived. Between 18th Street and Indiana Avenue on the old South Side main line a middle track operated for express service northbound in the morning and southbound in the afternoon was fitted with two-aspect electro-pneumatic operated semaphores.

---

[5] Optical bar code scanners to read reflective striped labels applied to the cars were tried first in this application. They were superseded by transponders, a device similar in function to Identra, but miniaturized and more sophisticated, reacting to a track-mounted interrogator coil.

Most other rear end protection blocks used three-aspect color light signals. In either system a train was detected by electrical track circuits and its presence made known to a following train through the wayside signals.

All five double track miles of the State Street subway, opened in 1943, were equipped with automatic block signals (ABS) to protect the rear of trains from the approach of a following train. This improved system has three-aspect color-light signals, each fitted with a left side mounted track trip to engage a spring-loaded carborne trip valve on any train attempting to pass a restrictive indication, causing an emergency stop by venting the brake pipe to atmosphere.[6]

To insure adequate braking distance under all conditions, circuitry was overlapped to protect the rear of a train by *two* stop signals, each enforced by a track trip. Thus, if a train had just passed one signal the second signal was still sufficiently far back to assure adequate braking distance for a following train.

This improved ABS system also utilized timing signals to limit speeds on grades and curves and to allow one train to slowly close in behind another standing in a station.

Under CTA the first major expansion of signalling came in 1951 when the four mile Milwaukee-Dearborn-Congress subway opened. It was similarly equipped with ABS, as, in the next few years, was the new line in the median of the Congress (today's Eisenhower) expressway. At about the same time the 1895-built elevated line from the north portal at Evergreen to Logan Square yard was retrofitted with ABS.

Typically, the wayside color-light ABS unit displays one of three aspects—

    Green   Proceed
    Amber   Proceed expecting to encounter
                    stop aspect at the next signal
    Red   Stop

If authorized to do so, the motorman may operate a track trip manual release lever on the wayside signal to cause the track trip to lower, and then proceed according to the rules, prepared to stop at any time. The ABS system utilizes insulated joints in one of the running rails to define the limits of track circuits, while the other rail returns the track circuit as well as the 600-volt propulsion current.

Speedometers and cab signals

Further expansion of signalling lagged for a few years. In 1964, the first speedometers were provided on some new cars. At almost the same time development of solid state electronics and audio frequency track circuits provided the technological breakthroughs that made available the continuous protection of automatic train control with cab signal and reduced the cost of completing the long needed signal system.

With ATC, the wayside equipment detects the presence of trains or other speed limiting conditions through track circuits powered by audio-frequency currents and sends through the rails commands which determine the maximum speed at which a following train may be operated. Equipment on that train compares this command with the actual speedometer reading and provides visible and audible displays to alert the driver. If the actual speed exceeds the command speed the motorman must reduce speed accordingly within a short time or the equipment will impose a penalty full stop.[7]

The pioneer installation of ATC was in 1965 on eight miles of the Lake route between Harlem Avenue and the Chicago River. It was installed on the Dan Ryan and Kennedy rapid transit extensions of 1969-1970. From 1975 on ATC was extended line by line to cover the remaining "dark" (unsignalized) trackage and to upgrade the segments that had signals without track trips. On portions of the North-South and West-Northwest routes trains operate under ATC-cab signal control, while under the remainder of those routes they are under ABS-wayside signal control.

The installation on the extension between Jefferson Park and O'Hare on the West-Northwest route includes CTC to permit train operation on either track in either direction, a feature which provides additional options for emergency operation as, for instance, in case of blockade of one of the tracks.

Grade crossings

Chicago's rail rapid transit is unique in having grade crossings, 26 of them, all in ATC territory.[8] They are equipped with conventional gates to span the right-hand approaching street lanes. The pedestrian crosswalks have gates or bells. There are automatic gate crossing signals facing approaching trains to indicate to motormen whether the gates are not lowered, are in the process of lowering, or are fully lowered.

---

[6] Side trips painted with whitewash have long since replaced center trips as standard. If passed *up* they are not destroyed, but mark the offending car.

[7] ATC is continuously in communication with the train, not just when passing a fixed point as with wayside signal and can therefore update its speed commands at any time if conditions ahead change. It monitors the continuity of both rails, not just one of them, and the use of audio frequency for the track circuits enables elimination of insulated joints in the track with a saving in maintenance cost and noise generation. The overall cost of ATC installation was estimated to be 40% less than ABS.

[8] There are ten grade crossings on the Douglas branch, six on the Ravenswood branch, two on the Evanston line, and eight, counting the pedestrian crossing in Dempster terminal, on the Skokie Swift.

Over the years the junctions at the northwest (Tower 18) and southeast (Tower 12) corners of the Loop have been changed several times to accommodate service routings. Since this photo, more changes have been completed here at Tower 12 to provide for the upcoming Southwest route to Midway Airport.

Van Buren/Wabash - 9-23-91 - A. Peterson

The two crossings in Wilmette on the Evanston line are also provided with woven wire gates, automatically operated from track circuits which swing horizontally across the tracks to constrain trespassing on the tracks. The controls are arranged to open only the gates covering the track on which a train is approaching.

Current developments and plans

Massive reconfiguration of the track layout at Howard Street is approaching completion in 1993. The project at this location, where the North-South, Evanston and Skokie Swift routes converge, includes rehabilitating the car storage yard and enhancing inspection shop facilities. Concurrently, the 1923-built electro-pneumatic interlocking is replaced by state-of-the-art "entrance-exit" technology utilizing a computer terminal as the work station for control of the plant. Completion of the Howard Street project will incidentally bring to the end of its life the 1899-built Wilson Avenue plant, last remaining mechanical interlocker in the Chicago rapid transit network.

Extensive track layout changes at Linden Avenue, the Evanston route terminal in Wilmette, are also under way, to be accompanied by replacement of the terminal station building. Also, for the future permanent south terminal of the Jackson Park branch at the Dorchester station currently under construction, a new interlocking plant will be provided.

Certainly the most impressive expansion and enhancement of CTA's signal system will come with the opening of the Southwest Rapid Transit Project now nearing completion between 18th & Federal and Midway Airport. There will be eight new interlockings along the new trackage plus new or modified ones on existing trackage. Included are 18th/Federal, 17th/Holden Court, Roosevelt/State and #12 Tower (Van Buren & Wabash).

Lots to look forward to...

◊ ◊ ◊ ◊

Chapter 9 - Maintaining the system

The total investment in fixed equipment and rolling stock of the CTA is well into the billions of dollars. The reasonable life expectancy of this property ranges from only a year or two for a supervisor's squad car to perhaps ninety years or more for a subway tunnel or steel elevated structure. After those times the property must be replaced or substantially reconstructed.[1] In the meantime, each item requires a certain level of normal maintenance throughout its life to keep it in a state of good repair. This chapter highlights what is done to maintain the system.

Maintenance is a major activity at CTA—its shops would rank among the larger manufacturing operations in the area. The annual cost of maintenance was approximately $237 million, or roughly one-third of the $721 million total operating cost for the year 1990.[2] For comparison, passenger service (predominantly the costs of drivers and their supervision) represents about one-half of the total cost and general administration, the remainder.

About 3,300 employees provide more than 2.6 million labor hours per year overcoming the ravages of wear and tear in keeping the plant and equipment clean, fueled and running in good shape.

---

[1] Assuming that technological advances don't make the property obsolete or that changing customer needs don't make replacement necessary sooner.

[2] All data based on FTA Section 15 Reports for calendar year 1990.

Initially it was necessary to make do with some streetcar barns for bus garaging, modifying minimally as at the 77/Vincennes operating station where hoists were added to service buses then used there—mostly 2300-series Brills. In March 1967 a purpose-built garage was opened at 78/Wentworth and the area shown here was again modified as a brake shop.

Modernization of housing had to keep pace with conversion from streetcars to motor and trolley buses. The number of operating stations was reduced, concentrating activity and conserving resources in the face of rapid loss of traffic to the postwar automobile boom. Within a few years, new garages were completed at Beverly, North Park and Forest Glen (shown). The latter opened 1955 to serve 350 buses for fifteen propane and trolley bus routes.

Within six months the wheels originally fitted to trolley bus poles were replaced by sliding shoes. Montrose and Irving Park streetcars were also fitted with trolley shoes. To improve operation and reduce wear, graphite grease was periodically applied to trolley wires by line trucks as shown here. This became unnecessary after carbon-insert trolley shoes were adopted in the period 1941-1945.

Elston route - 1933 - Krambles archive

The following indices illustrate the approximate relative weight of (**A**) transportation crews, (**B**) vehicle maintenance, (**C**) non-vehicle maintenance, and (**D**) other costs on (**E**) the average cost per vehicle-mile of bus and rail operation:

Mode	A	B	C	D	E
Bus	3.2	1.3	0.3	0.2	5.0
Rail	1.6	0.9	1.0	0.5	4.0

Maintaining Buses

Bus maintenance includes frequent running inspections, cleaning and servicing, carried out in nine garages. Heavy repair work, such as engine overhaul, body damage rebuilding and painting, is concentrated at South Shops, 79th/Vincennes. Nearly 74 million bus miles were operated between the almost 2,200 vehicles of the fleet in 1990. At that time the average age of the buses was about 14 years; 68% of them were from 12 to 25 years old.[3] All told bus maintenance uses about 1,800 employees, or about one person for each vehicle on the street during peak service.

As is mentioned elsewhere, there has been a continuous effort since 1947 to upgrade the buildings and tools available for bus maintenance, yet after 45 years there are still garages in daily service that were built more than 80 years ago as streetcar barns. Of course, all have been revamped and retooled, but the old layouts are far from ideal for an efficient, environmentally desirable operation. A garage replacement program will probably remain high in priority until all have been brought up to date. Current concepts are designed to enclose all activity so that circulation for cleaning, washing, fueling, inspection, repair and storage are conducted in weather-free, pollution controlled, noise absorbing, easily supervised indoor surroundings.

Bus maintenance includes a modern computerized Vehicle Maintenance System to facilitate retrieval or field repair of any bus that may develop a defect requiring a road call.[4] As one example of how this works, an operator experiencing a problem that disables a bus reports it by radio. The control center relays it through the VMS, producing a hard copy report at the home garage. The foreman at that garage gives a copy of this report to the service person who is assigned. This person, a qualified mechanic, takes a replacement bus, driving it to a rendezvous with the disabled one. The operator takes the replacement into service. The mechanic corrects the problem and/or brings the crippled bus back to the home garage.

During busy rush hours in the central part of the system there are roving mechanics working out of vans who similarly meet buses with problems and correct minor difficulties (such as overheating or windshield wiper failure) so that service delays to riders are minimized.

Around the clock there are heavy duty wreck trucks that can be summoned by radio to clear a street accident, to move a bus that has a dead motor or a flat tire, or even to pitch in on solving some kinds of rapid transit problems. During a normal business day three such trucks are deployed to cover the service area. In addition, a supervising chauffeur, working from a van and available around the clock, has special responsibilities to work with the fire and police departments in coordinating efforts dealing with problems which may delay CTA services.

---

[3] FTA accepts 12 years as a reasonable life cycle for buses.

[4] There were more than 27,000 bus road calls reported in 1990.

Of course, taking care of the bus system includes a lot more property than just the buses and the many trucks that support bus operations. To mention just a few kinds of things in addition to buses and garages, there are thousands of bus stop passenger information signs and shelters. There are more than a hundred off-street turnarounds at the ends of lines. These involve varying amounts of heating, ventilating, lighting, draining, repair of vandalism, etc., etc.

Communications equipment for buses also needs attention. Each bus has a radio and there is related wayside equipment for vehicle location as well as voice communication. An appropriately complex control center with its base transmitting and satellite receiving stations is provided to support it. Plenty to take care of...

78th/Vincennes yard - 2-26-76 - A. Peterson

One of the functions of the Utility Department is to provide round-the-clock wreck truck service to tow disabled buses or to clear street accidents that are delaying service. Chauffeur Howard McMillan shows off Diamond-T truck 210 which works the south side. His department operates a sizeable fleet, with a wide variety of vehicles for transporting material, ranging from heavy steel to bulk money.

Minimizing the use of poles and overhead cable for distribution of power was a practice of both CSL and CRT. This meant miles of underground ductwork, with hundreds of manholes, linking more than 45 substations or circuit breaker houses. This invisible network carries power, control and communication cable, a significant maintenance burden. Here, hi-rail truck 259 and crew are working on cable linking Sedgwick substation to the 'L' and subway.

Mohawk/North - 8-28-91 - A. Peterson

West of Cicero/Lake - 8-22-92 - G. Krambles

Cicero/Lake 10-29-91 - G. Krambles

Principal maintenance facilities

🚆 Rail vehicle maintenance
🚌 Bus maintenance
■ Plant/utility maintenance

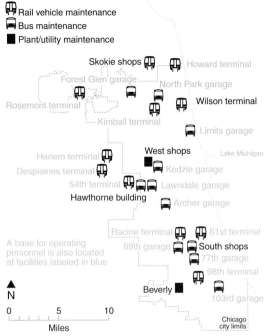

Map: Roy G. Benedict Publishers' Services

Chicago's 'L' structures include more than 12,000 columns. They rest on masonry or reinforced concrete spread footings buried in the ground. After 90 to 100 years these have begun to crumble causing pumping of the columns when a train passes. Rebuilding at the rate of 1,000 footings per year has been in progress now for three years.

Between Rockwell and Laramie on the Lake line construction costs were saved by locating support columns in the street. At other places, street widening has extended roadways around all sides of columns. Little by little these obstructions to street traffic are being eliminated from critical locations.

Maintaining Rapid Transit

Rail vehicle maintenance similarly includes frequent running inspections, and cleaning and carried out in the terminals. Heavy repair work is concentrated at Skokie Shops, 3901 Oakton Street in the Village of Skokie.

Nearly 57 million car miles were accumulated over the 1200+ rail vehicles in 1990. The average age in that year was less than 14, having just been reduced substantially by the acquisition of 600 cars of the 2600-series.[5]

Rail maintenance requires the equivalent of about 1,760 people, or a little less than two persons for each vehicle in service during the peak. The staffing is divided about 3/4 to vehicle and 1/4 non-vehicle maintenance, although the cost is about equally divided between these functions, due to contract work and materials.

---

[5] Even including 5-50- and 6000-series cars, remanufactured from 1946 street cars, this was major improvement: by 1947 rapid transit cars often soldiered on beyond a half-century.

Structure sandblasting and painting is a perpetual burden of maintaining 199 track-miles of 'L' structure. Today, environmental reasons demand enclosure of the work area and other procedures to guard workers and neighborhood while assuring results essential to extend the life of steel.

The short segment of the South Side elevated on Harrison Street between Wabash Avenue and Holden Court has changed little in its 95 year life. It contains the sharpest curves in the system at 89 ft. radius, protected by guard girders against possible but highly unlikely repetition of the 1977 derailment that occurred on the similar curve at Lake/Wabash. *(see photo page 116)*

Montana/Sheffield - 8-15-91 - A. Peterson

Wabash/Harrison - 5-22-80 - A. Peterson

Nelson/Sheffield - 4-9-92 - A. Peterson

Because rail cars last much much longer than buses, the main shops at Skokie are loaded with some kinds of heavy repair work not encountered for buses at South Shops. Thus, while Chicago's cars rarely encounter damage repair from fender-bender collisions at the 25 road crossings as compared to its buses, which must work in mixed traffic over almost 1,500 miles of city streets, rapid transit shops must be capable of tear-down and complete rebuilding of all major components, usually at least twice during the life of each vehicle.

Exterior and interior painting was historically a major work load for Skokie shops, as it remains today at South Shops for buses. However, in recent years epoxy finishes more than doubled the life of a paint job and the adoption of

Merchandise Mart/Ravenswood station - 7-2-88 - A. Peterson

Work trains distribute material that can't be handled by motor trucks. Since 1975, motive power is all-electric (ex-6000 series) cars equipped with cab signal ATC protection. Trailers for the lading are moved in the middle of the train so there is a driving cab available at each end. Trailers have compressors and air brakes controlled by the digital trainline. The scheme was developed by MCERA John F. Humiston during his CTA career as an equipment engineer. The train shown, led by heritage unit 6101-6102, is transferring components between Skokie and terminal shops.

The third station on this site, once the terminal of the Chicago & North Western Railway, was built without interrupting service. Privatization here is represented in the design and construction. The job included a new bridge over the one remaining track of the C&NW on Carroll Avenue (at lower left), a route considered for the Central Area Circulator.

stainless steel car bodies lined with permanently finished fire-resistant plastics since 1969 has further reduced this task.

The buildings and tools taken over by CTA from CRT were an immediate target for improvement. The most modern facility acquired was the main shop at Skokie, the newest portion of which dated to 1930. Each of the four underlying divisions (Met, North, South and Lake) had its own main shop dating from the 1890s.

Skokie was only equipped to do painting and car body work, then chiefly carpentry. Heavy tools, scattered over the five facilities, basically were of the same vintage as the buildings. One of the first moves was to concentrate all main shop work at Skokie, relocating crafts union personnel and the best of their existing tools there. Over the years since, tooling has been constantly updated to deal with the changes in technologies.

CTA has pushed the transition from its predecessors' heavy dependence on carpentry and blacksmithing to aluminum and stainless steel fabrication and welding, and from the slow and

Tracks wear out after a half-century. Rehabilitating them presents the dilemma of whether inconvenience to the public is lessened and construction costs are minimized by a series of piecemeal renewals done under traffic (with single-tracking in off-peak hours), or, by shutting service down totally for many weeks to do massive renewals. Rehabilitation in the State Street subway (North-South route) in 1991 was done by suspending weekend service via subway and rerouting trains over the 'L'. Upper photo shows stockpile of ribbons of welded rail along the connection (nearing completion) to the Dan Ryan line; bottom is subway installation in progress.

18th/Wentworth/North-South route - 4-21-91 - A. Peterson

Clybourn station/North-South route - 4-21-91 - A. Peterson

Built in 1899 by Northwestern Elevated Railroad, Wilson shop, North-South route, at this writing is serving its fourth generation of cars. Cramped and ill suited to the performance of some maintenance on sophisticated modern cars, it is to be replaced by the thoroughly modern shop being built at Howard yard.

ponderous wheel turning that required wheel and axle sets to be disassembled from trucks to a modern computerized underfloor wheel truing machine that, in just a few hours, can reprofile all wheels to more exacting standards than ever before and do the work right in place on a car. Propulsion control, signalling and auxiliary system packages have progressed from simple but crude electro-mechanical devices to sophisticated electronic and microprocessor technology.

Inspection work at terminals was mostly done in antiquated buildings poorly located in yards not always at ideal sites with respect to the operating requirements of the services. Heating, lighting and working space around cars was at best inadequate and at worst, as at the Lake Street shop, with its dirt floor, deplorable. At places such as Howard Street and Linden Avenue, the pits were in unsheltered yard tracks, unbearably hot in summer and impossibly cold in winter. Car washing was done outside by hand, using long handled brushes, but obviously only when weather permitted.

Over the CTA years much progress has been made to correct these shortcomings, with replacement facilities purposefully designed for efficient performance of inspection and trouble-shooting tasks on state-of-the-art rolling stock. Such modern shops have replaced the old at almost all terminal locations, but work at 61st Street and at Wilson-Montrose continues in the original buildings, now more than 90 years old![6]

Substantial advances in car washing have been made by installation of semi-automated machinery, but in Chicago's long and tough winter season, washing of rail vehicles is much more difficult to fit to a preprogrammed schedule than it is with buses, where all-weather indoor washing is normal today.[7] More nearly ideal solution of the problems of car washing and interior cleaning remain for the future.

For the rail system, maintenance of physical plant other than the vehicles provides a major workload. Included are tracks, structures, power distribution, signalling, communications and control center, vehicle maintenance buildings, tools for fabrication, assembly and repair, and, a wide variety of on and off road utility vehicles.

---

[6] A replacement facility is currently under construction at Howard Street, a strategic location improvement.

[7] Efficiency of rail vehicle washing is proportionate to the length of a cut of cars that the washer can accommodate, but in Chicago's relatively cramped yards it's rare to find enough space for a building for all-weather washing plus trackage to facilitate handling full length trains between storage tracks, through the washer and back to storage without impairing line operation.

In one case, to clear the way to build a terminal car inspection shop adequate for today's needs a small temporary facility was first built so that the then-existing shop could be demolished and the new permanent one built in its place. Actually, even the temporary unit, shown here, had proved a substantial improvement compared to the frame structure that had served the Ravenswood line since 1907.

Primary base for this activity is the impressive West Shops, at 3900 west, stretching from the Chicago & North Western right-of-way on the north across Lake Street and south to Maypole Avenue. Based here are the departments responsible for track, structure, power distribution, communications, building and utility services. Crafts persons and managers required for these activities work from the West Shops. Here are major tools for fabrication and assembly of all manner of standard or "one-off" components required for power, way, buildings and grounds maintenance and specialized construction.

Of course, practically all facilities maintenance and plant renewal work is destined for sites scattered system-wide, and much of it must be performed where used. However, many complete components are prefabricated efficiently under optimal working conditions at the West Shops. At least partially completed, they are then transported to their permanent field location for installation. This insures rapid installation and minimal disturbance to transit service and community life. Examples include track panels, turnouts and crossings, agents' and supervisors' booths, 'L' station stairways, structure columns and stringers, circuit breaker panels, and bus stop shelters.

An adjunct to every garage, terminal and each of the main shops is a storeroom where an adequate stock of spare parts and materials is securely banked for ready retrieval when needed. Because of the ever increasing value of supplies, continuing effort is made to minimize the stock sitting idle and yet to be sure that equipment does not remain unavailable for lack of parts. The goal is to have stock at the counter "just in time" and perfecting a management information system to do this has a high priority.

There are myriad locations where there are concentrations of maintenance activity that justify local facilities, ranging down to an electrician's closet for storage of spare lamps or the very small janitor's closet built into one end of a modern agents' booth.

These few paragraphs are a very much condensed summary of the effort required and the facilities provided to keep the multi-modal transit system of CTA moving. Forces of wear-and-tear, vandalism, mother nature, human frailty and obsolescence are at work every day tending to degrade transit system performance. It is the vital responsibility of maintenance to keep transit's tools and equipment in good repair.

◊ ◊ ◊ ◊

Chapter 10 - Service control

A little background

A primary objective of CTA is to provide on-time scheduled transportation service. However, many unpredictable and uncontrollable factors, such as weather, traffic congestion and equipment failure can disrupt operation, causing delay to riders and overcrowding of vehicles. To deal with service problems effectively, all available information on current operating conditions, complete schedules for the entire system, together with appropriate remote control equipment and especially with extensive communication equipment has been assembled by CTA into an Operation Control Center. Staffed with skilled and specially trained controllers, it is in a centralized location convenient to the general offices.

It was not always thus. Unification of streetcar and rapid transit into a single management brought with it the need for coordination of service control. Postwar street traffic conditions, constantly worsening with the meteoric increase in automobile population, aggravated irregularities in surface transit. At the same time, stiff competition from the private auto bled ridership.

Service objectives for both surface lines and rapid transit were changed from the historic approach of competition between them to making each mode do what it was best fitted for, maximizing both coordination and productivity along the way. For example, universal transfer between surface and rapid transit was established, setting the stage for new operating plans. In just a few years, to improve the speed of rapid transit, 99 little-used stations were eliminated and all-express "A and B" service was instituted, while long through-routed streetcar lines *paralleling* rapid transit routes were separated into shorter routes *feeding* 'L' stations. The goal: **integration,** not **duplication.**

The way it was

To manage the emerging integrated streetcar, bus and rapid transit system, a centralized service control function was foreseen in 1948. It was to be equipped with innovative tools utilizing leading edge technology of the time. It was to update accepted practice in the industry at the time, which was essentially to leave the prevention and overcoming of irregularities to operating crews and local supervisors, ill-equipped to take a global approach to such problems. Commonly little or no action was taken, lest it worsen rather than mitigate a problem. Inadequate information made effective correction impractical.

CSL did the best it could with a supervision system of about forty radio-equipped squad cars, plus trucks capable of heavy tasks such as towing. There was also a network of "point" supervisors, working at key locations equipped with dial phones. There were other

Edison Building, Chicago - 1941 - G. Krambles

Junior engineer George Krambles works a wartime shift as power supervisor for the Chicago Rapid Transit Company, controlling 37 substations, electrical switchgear and over 200 miles of cable connections as well as dispatching electrical equipment repair employees. It was the beginning of today's coordinated control center.

The line supervisors' office was developed in the 1950s to centralize control and communication of the rail lines. Automatic train despatching (left) was provided to start trains precisely at every terminal and recorder charts (right) made real-time permanent records of schedule adherence. It became the rail portion of today's operation control center. Controllers shown (left to right): Edgar Ferguson, Walter Pavoni, Ollie Winston and Tommy Hughes.

phones mounted on light poles throughout the city. Focal point was a "radio-telephone room" in the main office, manned by dispatchers.

A streetcar motorman, conductor or operator (driver) with a problem had to find a phone, stop and call in. The dispatcher who received the message then relayed it (by phone or radio) to appropriate personnel in the field for corrective action. Much time was lost in the process and many problems went without attention. It was not uncommon for a bus to be lost to service for hours.

Service supervision on CRT consisted of dispatchers or trainmasters at terminals or key locations along the lines. They could communicate with one another to a limited extent, using the private dial phone system that extended throughout the CRT property.

There was a central power supervisors' office for managing the (then-)37 power substations and the electrical distribution network of the rapid transit. While this office had no authority over train operation, by default it became the focal point for management information in major emergency situations.

On both the surface and rapid transit systems, the net effect of this was that irregularity of service tended to be the rule rather than the exception. Dealing with major problems was especially frustrating with so little command and control capability, almost none of which was centralized at some point where an overall corrective strategy could be formulated and directed.

CTA General Manager Walter J. McCarter directed his Staff Engineer's Office (the operation planning department, then headed by Louis M. Traiser) to conceive an approach to meet developing service control needs and to manage initial implementation. This began in 1948 with a small test on a short segment of the Lake rapid transit route. By the early 1960s, it extended to all of the rapid transit.

Experiments with bus location monitoring and radio communication began almost simulta-

A corner of the operation control center. In the foreground are positions (momentarily unoccupied) for system controllers with consoles for radio channels dedicated primarily to dealing with bus supervisors, wreck trucks and operations managers. Rear left with the many colored lights, is the power control area where work is done largely with direct remote control or by PBX telephone. At far right, barely in this view, the rail control area, where event recorder charts show train schedule performance in real time, and communication is by other dedicated radio channels linking train operating and supervisory personnel, as well as to the public by loudspeakers on station platforms. Out of view to the right are the bus controllers who work directly with bus operators, using still more channels. The operator at console marked 3 distributes equipment trouble reports to the garage that must repair or replace the bus. *(See photo, page 89)*

neously with Ralph W. Tracy as project manager. Thus, CTA pioneered in upgrading service control in both the rail rapid transit and bus transit fields.

Service control on rapid transit

When the development of rapid transit centralized service control was begun in 1948 about 65% of the track in service was not equipped with block signals. There were spacing boards mounted between tracks which were a guide to the minimum safe distance to be maintained between trains. In such territory, safety depended on motormen avoiding rear-end collision by operating strictly according to "on-sight" driving rules. Compliance was surprisingly good, but sadly, not perfect.

Unlike the rapid transit systems of other cities worldwide, which were fully block-signalled, it was generally not possible in Chicago to generate a central office display of train locations from track circuits already in existence; there were no overall track diagram boards showing, with miniature colored lights, where each train was.

Starting with a "clean slate", a unique train chart approach was originally chosen to minimize the time and cost of initial implementation. In a line supervision area of the operation control center it produces an analog display showing clearly and simply where train performance is good, bad, or so-so. A controller, who has advanced from the practical background of train service, quickly learns to spot developing irregularities from the train charts and gets a prompt feedback of the effectiveness of whatever service correction measures are used. The charts become permanent records and are stored for subsequent retrieval for post mortem review of strategies or for complaint or claim analysis.

The charts were supplemented by automatic train dispatching (programming) devices connected to train terminals and other field locations by telephone lines to display appropriate starting signals precisely at the time when each train is to start a trip.[1] There are pushbuttons, displays and other controls for modifying the dispatching, and for operating the communications system consisting of radio, telephone, and station loudspeakers.

---

[1] In some cases, the automatic train dispatching signal also initiates the operation of an unattended interlocking plant at the remote terminal to initiate the request for a track path so it will be ready for the departing train.

The first application of CTA's venture into centralized rapid transit service control was in 1951. Train indications and platform loudspeakers were installed in a short section of the North-South route covering the State Street subway in the central business district, where passenger loadings were the highest in the system, at that time actually running **greater than** full system capacity for the peak of the rush hour. Here the urgency of effective corrective action to overcome irregularities was the greatest, as a delay of only a couple of minutes (one train headway) would immediately cause overloading which all too often would mushroom out to a 20- or 30-minute gap between trains, reverberating up and down the line until well past the rush period. Success with this trial effort led to gradual expansion of the control system, then called **line supervision,** to blanket the whole rapid transit network.

Concurrent with this expansion, the **train phone** was developed to provide the essential communication needed to exchange service correction and trouble-shooting instructions between OCC and the driving cab of trains. While this technology, which superimposed a radio frequency signal on the third rail, was a step forward, it was not dependable under all circumstances and was replaced when ultra-high-frequency radio came on the market in the early 1980s.[2]

Nowadays, each train service employee carries a portable walkie-talkie radio monitoring a distinctive channel according to route. To assure maximum knowledge and to minimize the number of messages and the time needed to inform those concerned of what's going on, all radios on each channel are in party line repeating mode. The use of portable radios has the advantage of maintaining capability of com-munication even when it is necessary for an employee to be away from the normal working location, e.g., out of the driving cab to do trouble-shooting or to tend to a disabled passenger.

A modern dial phone system reaching fixed locations throughout the Authority is used for a tremendous volume of message traffic that does not require talking with crews on the trains themselves. For messages to the public at stations, there is a comprehensive system of loudspeakers covering every platform, arranged to be operable by an employee at the scene or by remote control from the OCC. All passenger cars have loudspeakers which can be operated by the train crew. Tests have been made of operating these by remote control from the OCC, but experience has shown better results by having train announcements made by the train crew.

---

[2] There was dramatic proof of the value of radio in dealing with operations in the winter of 1978-1979 (see Chapter 11), when Motorola, on short notice, supplied a number of 450-megaHertz walkie-talkie transceivers, lifesavers without which rail operation would have come to a halt.

The presence of alternate routes through the central area to the north and south via 'L' and subway, has proven its worth operationally. For example, earlier on this morning a northbound work train accidentally knocked over several carlengths of third rail in the State Street subway. The damage was repaired by early afternoon, but in the meantime, northbound service was rerouted over the 'L', a reroute coordinated by the control center.

Lake/Wabash - 9-26-90 - A. Peterson

Howard/Hermitage - 1962 - Krambles archive

Integration of Chicago's rail and bus systems led to extensive investment in interchange stations—at one of the most important, three rail lines and eight bus lines exchange riders. Commitment to coordination is supported by the unified rail-bus-power operation-control-and-communication center at the Merchandise Mart.
In the 30 years since this aerial photo was made improvements have continued and even today further major changes are expected to increase the efficiency, safety and comfort with which the public is served.

Service control on buses

Improvement of service control on buses has lagged that for rail. Although the general principles are similar for both modes, more accurate operation and hence, closer control has been developed for rail, because:
- √ - a train delayed may inconvenience up to fifteen times as many people as would a bus,
- √ - stoppage of one train blocks trains following on the same track (and often even on adjacent tracks), while a delayed motor bus can usually be passed by its followers,
- √ - service stoppage between stations on elevated structure or in subway has the potential to alarm passengers, while an alarmed passenger on a bus stopped on a city street for an inordinately long time can walk away, and
- √ - long delay in train service during rush periods cascades into a breakdown of connecting bus service if traffic load is deflected to alternative bus routes as well as by the heavy rate at which bunched trains may deliver passengers to buses at transfer points.

That's not to say that buses don't need a sophisticated service control system; operation in mixed traffic in the city's streets introduces many irregularities that never occur in the private right-of-way of the rail system. But, on the other hand, a missed trip in a rush period due to a disabled vehicle is obviously much more likely among the 1,900 independently operated buses out on the street than is the case among the ±127 trains even though they need about 923 cars.[3] Unlike the rapid transit situation, a bus delay going unnoticed or a bus getting lost are possibilities for which there is presently no positive detection.

At the bottom line, the development of a bus service control system approaching the effectiveness of that for rail had to wait for a dependable technology for bus communication and location, generically termed *Automatic*

---

[3] A disabled bus must be replaced with another, usually brought from the garage. A disabled rail car may cause a delay, but except for Skokie Swift, all trains have more than one car (many in rush hours have eight) so that a car with a problem can usually be cut out and the train will continue after only a brief delay.

At each garage a computer work station reproduces requests logged at the control center concerning bus defects reported as soon as noticed by bus operators (via their radios), together with a rendezvous time and location assigned by the bus controller. Hard copies of the reports are then assigned to mechanics to expedite repair or replacement with minimal disruption of passenger service. A similar system, modified to fit the differing needs of train operation, is also available for the rail controllers to alert the terminal and yard of car defects reported (via their radios) by motormen and conductors while en route. Unlike the buses, which in worst case may be *curbed* to await replacement, every effort is made to keep trains moving, and therefore train crews and their supervisors are extensively trained to do basic trouble shooting under guidance of the rail controllers. *(See photo, page 86)*

North Avenue garage - May 1977 - G. Krambles

Vehicle Monitoring. As far back as 1935 the streetcar system had tested an automatic headway recording device consisting of a contact clip mounted adjacent to the overhead wire in such a way as to receive an impulse each time the trolley of a streetcar passed. This was fed into a pole-mounted chart record which could be removed the next day and taken to the office for analysis. Interesting, but hardly useful for real time action.

CTA began research into a better solution for that problem in the 1950s and by 1958 was testing the use of a very low power radio transmitter on a bus to trigger a signal in a wayside receiver which was relayed by telephone line to the central radio room. The next step was to encode the signal so as to identify ***the*** passing bus, not just ***a*** passing bus.

During the following ten years the studies and tests continued, leading to equipping 500 buses with UHF two-way radios having the capability of reporting bus location with respect to discrete "signpost" transmitters strategically located at selected timepoints. A significant new feature introduced at this stage was the provision of an emergency alarm, or, as it became known, a "silent alarm," by means of which a bus driver, without saying a word, could initiate a distress call for police assistance. Gradual reiteration of the hardware and software and expansion of the system with more signposts improved accuracy and reliability of AVM and by 1978 all 2,400 buses then in the fleet had the newest radios.

Even with the communication problem largely licked, the development of adequate capability for overcoming a delay once it occurs has stubbornly resisted solution; all the while traffic congestion gets worse. The cost of adding manual supervision is prohibitive given such questionable effectiveness. Strategies that work well on the rail system with its private right-of-way are impractical for buses. At the time of this writing, the pursuit of better answers continues. In connection with long overdue modifications to CTA's basically 40-year-old general office space (which may require relocation of the OCC in any case) and renewal of the aging bus radio equipment, a fresh look is being taken at application of recent developments in microprocessor and communication technologies in America and abroad to the long sought AVM goal.

◊ ◊ ◊ ◊

One burden of the days of its predecessors that CTA has been relieved of is snow removal from the streets on which it operates, originally one of the *quid pro quos* of a street railway franchise. Here a CSL sweeper was working southbound clearing a path on Clark Street passing the Chicago Historical Society.

Chapter 11 - Battling Mother Nature

All extremes of nature present operating problems to transit. Transit riders may become impatient and less careful in boarding or alighting. Some pedestrians and motorists take imprudent risks in seizing the right-of-way.

Heat and cold challenge bus engine, air cooling and heating performance, increasing geometrically the need for road calls to rescue disabled buses and for service control action to smooth out resulting irregularity in headways. Rain may affect adhesion, sometimes in an unpredictable manner, resulting in slow operation and fender-bender accidents. High water may block routes, particularly at underpasses, a common problem for bus operation on the south and southwest sides of the city.

High wind makes steering more erratic and may block routes with debris. Fog, cloudburst rainfall or heavy snowfall occasionally reduce the range of visibility to as little as 75 feet, constraining operating speed to insure safe braking distance within the range of sight. Since the implementation of a modern signal system with automatic train stop or, latterly, automatic train control with cab signals, the potential hazards of low visibility on train operation have been greatly reduced, but there remain risks when visibility is reduced as at stations and grade crossings.

Earthquakes, disastrous for some west coast transit systems, fortunately up to now have been no problem in the CTA area.

In Chicago, all things considered, Mother Nature's greatest threat to good operating performance of bus and rail transit is winter's triple-whammy of snow, sleet and ice. CTA must prepare in anticipation of the possibility of **all kinds** of weather interference, but exactly what does **all kinds** include?

Having over a century of transit operating experience provides useful criteria, but in overcoming weather problems adequate preparation followed by effective corrective action remains at times more of an art than a science. The details of extremes of weather are not precisely repetitive. Small variations in such factors as temperature, wind direction, time of day and duration of precipitation (and, of course, in the combination of those factors) can make the difference between smoothly maintained service and complete frustrating paralysis.

To assist in dealing with weather-related problems, CTA supplements official U.S. Weather Bureau services by retaining a consulting meteorologist to alert it with forecasting pinpointed to the Chicago area, even down to a specific neighborhood when necessary, and updated in real time.

Winter on the streetcar and bus system

Mobilization for snow removal has undergone a metamorphosis over the CTA years. This is related to the growth of the automobile population as well as to the change of transit mode from streetcar to bus. Historically, when the streetcar lines were built they took on the responsibility for removal of snow from their tracks.[1] A large fleet of special purpose cars to sweep, plow and salt the tracks was built up.

When temperature coincides with dew point, pea soup fog reduces visibility below braking distance for normal operating speeds and at the same time it brings extra riders to the trains. A major safety enhancement under CTA has been provision of rear end signal protection over what was the world's last rapid transit network without it. Until this was done special "fog orders" were issued. With today's signalling, fog rarely delays service.

Later, as streets throughout the area were paved and automobiles began to replace horse-drawn wagons, municipalities began to develop capability for snow removal. Where buses were introduced, the transit system continued to be the prime actor and applied skills developed from streetcar experience to truck mounted snowplows and salt spreading vehicles.

Over the last twenty years, as auto and truck usage continued to grow, responsibility for snow removal from the area's streets (most of which are *not* used by transit services) gradually became the prime responsibility of the highway agencies. Transit remains a partner in the work, with its representatives participating at the City's snow command center during storms to coordinate the work, assuring priority to clearance of streets used by transit and helping identify trouble spots. CTA, of course, also has vast areas of bus turnarounds, garages and parking lots which it continues to clear for itself.

Winter on the rapid transit system

For the first eighty years of their existence the rapid transit rail lines operated almost free of interference from snowfall. The steel 'L' structures are practically self cleaning of snow falling at moderate rates and accumulating up to 8"–10". Even in heavier snowfall it is normally only necessary to maintain some extra train and car service through the night to keep

---

[1] In some cases this duty, along with sprinkling of unpaved streets used by car lines, was a requirement imposed by a municipality in consideration of granting a franchise.

Montrose/Kenmore - 1946 - G. Krambles

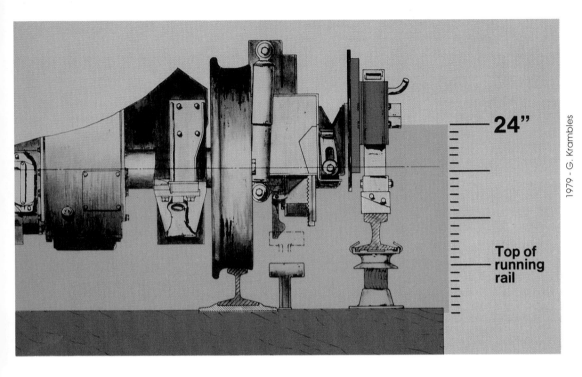

Sketch showing a cross-section of a typical truck superimposed against the accumulated 24" depth of hard-packed snow (including a fresh topping of 20" of which had just fallen) confronting trains by January 14, 1979. Note that it was about a foot over the third rail. Motors were obviously ill-suited to the job of plowing snow. Today, all cars have snow-squeegees which can take care of fairly heavy single storms by pushing snow from the track area to each side. But if the tracks and the space each side of them should ever again be already covered with a glazed crust several inches deep even with those there would be the problem of where to push the snow. Successful removal under such circumstances is dependent on thawing cycles between storms, and such did not occur in the 1978-1979 winter!

the lines open. More about snow fighting later.

However, the 'L' is much more vulnerable to icing problems when rain falls with temperatures hovering near the freezing mark. Ice is an excellent electrical insulator and its presence at contact surfaces will cause severe arcing and even total loss of traction power. If the temperature has been above freezing and then falls abruptly to a point just below 31° ice begins to shroud trolley wires. If instead it begins to rain when the temperature is rising after a long very cold spell, the relatively small mass of trolley wires will allow them to follow air temperature closely and not ice badly, whereas the specific heat of the massive contact rail freezes mist and rain into hard ice even when air temperature is above 32°. The impracticality of having a cover that would clear the dynamic outline of the cars over a third rail only 20" outside of the gage line forced the development of mechanical sleet cutters within the first decade of Chicago's rapid transit operation.

Under CTA, vulnerability to service interruption due to ice storms has been reduced. Almost as soon as CTA took over, the propulsion control of 4000-type cars was changed to work from the car batteries, uninterruptible as compared to the previous use of 600-volt DC power obtained the third rail shoes of the first car of a train. Next, trolley wire was eliminated from the Lake and Evanston routes and now remains only on half of the Skokie Swift route. Under capital grant programs, miles of third rail heater cable under remote radio control has been installed on upgrades and at key locations to raise rail temperature the small amount required to mitigate the problem.

The Winter of 1978-1979

CTA's most trying winter, that of 1978-79, was so unusual as to warrant detailed description here.

According to records of the U. S. Department of Commerce Weather Bureau, Chicago's total annual snowfall averages around 33" and seldom exceeds 40", accounting for less than one-half the total wintertime precipitation. Up until the 1978-79 winter, their records back to 1899 show only a few seasons where the total snowfall exceeded 60":

1917-18, 64" of which 42.5" fell in December
1951-52, 66" of which 33.3" fell in December
1966-67, 68" of which 28.9" fell in January
1969-70, 77" of which 20.2" fell in December

But in 1978-79, the season total was 88.4". There was compacting but hardly any melting: By the end of January there was an accumulation on the ground of more than 47" and it was mostly solid ice!

Most of the time it was very cold, with a record setting temperature of –19° F one day. During all of January, the temperature went above freezing only on three days and then only for a few hours. February began bitter cold and had no temperatures above freezing until the 20th. Real thawing weather finally arrived about March 18th!

Just a few notes from the operations log tell the story:

Four severe ice storms in December damaged rapid transit motors and controls. Normally there would be no more than one ice storm every other winter.

On December 31 a snowfall of 7.6" filled tracks in yards and surface areas up to the top of rail. Very cold weather in early January allowed no thawing.

On Saturday, January 13 there was a snowfall of 20.3". All trains were running with extra cars, but when snow depth accumulated up to the motors, it stopped the Skokie Swift, Evanston and the outer ends of the Ravenswood, Douglas, Congress, Lake and the North-South beyond Wilson.

Record cold (–19° F) on January 14 and 15 prevented effective snow fighting, while attempts to keep trains moving only packed snow into rock-hard ice. By the 15th, some 36 miles of track was covered by 36" to 48" of snow drifts. Nevertheless, by January 18th enough of this was cleared to restore some service on all lines.

Then, on January 23 an additional 8" of snow fell. Because the accumulation of ice between the rails could not be removed, this modest new fall again stopped operation on most ground level and embankment tracks.

Even after these were cleared, snowfall on January 27th, this time only 4", threatened to close the tracks, which by now looked like trenches, because there was no where for the new snow to go except across the rails.

Snow fighting technology develops

Over the years some special facilities had been put in place to help keep trains moving in winter storms. Cars were fitted with truck-mounted "sleet scraper" blades that could be lowered manually to scrape snow and ice from the head of the contact rail.

Its distinctive International Orange belt rail stripe identifies car #24 as Superintendent (and MCERA) Bruce Anderson drives a snowfighter extra against continuous three-foot drifts. Accepting unbelievable punishment to mechanical and electrical systems was part of the price paid in an unrelenting effort to keep rail lines open against the effects of a streak of 100-year winter storms.

Davis/Evanston station - 1/16/79 - CTA

In addition, some cars were fitted with small plows suitable for dealing with the normally expected maximum of 8" to 12" snowfall. As soon as it was evident that there would be more snow than these could handle (and no one had predicted the combination of 28" of snow with bitter cold) the shops manufactured several sets of heavy plywood plows capable of pushing snow even if it accumulated up to the car floors... which it did!

Plywood, laid in several cross-grained layers up to 6" thick, was used to provide strength with the electrical insulation needed to work near the 600 v DC third rail. These wooden plows were each attached to one end of a standard married pair passenger unit, which when coupled into a long train provided a snow pusher of up to 4000 hp.[2]

Massive damage to cars

In the effort to keep all lines open, trains were kept out on the lines attempting to push their way through, even though this meant that some of the snow *actually was being plowed by the car motors and undercarriage*. Not surprisingly, snow was forced into parts of the equipment that are normally a foot or more above the top of heavy snow accumulation.

The damage this caused to brakes, motors, generators and couplers plagued CTA for months. The snow and ice problem was exacerbated on the routes operating in expressway medians by the liberal quantities of salt applied in clearing the highway lanes. Salt, a well-known enemy of electrical insulation, was shorting third rail circuits and causing equipment fires.

This problem was particularly troublesome on the long Dan Ryan line. Each day the car casualty list increased more rapidly than repairs could be made. In a futile attempt to keep the Dan Ryan line going, more cars, including the newest ones, were borrowed from the other routes but the salt and ice problem there just would not be overcome. By Monday, January 29th, because of the daily decrease in the number of OK cars available for service, more than half of the 890 cars needed to provide a normal weekday rush period service over the system were disabled and could not be used.[3]

With so little capacity available, there was severe overcrowding. Trains were being stalled by passengers trying to force their way in and out of fully loaded cars. People were riding outside between the cars or at the rear of the train; some were trying to climb in at the end of cars, even jumping on and off moving trains. All over the system, the fewer than normal available trains were not only attempting to carry regular riders, but also the many foul weather patrons who normally would drive.

On the Dan Ryan line, northbound trains were receiving riders to crush capacity right at the terminal at 95th Street. Stopping such trains with no space for more people at intermediate stations only worsened the delay and compounded the problem. It was taking 30 to 40 minutes to make a run normally only 10 minutes long. To make all the intermediate stops in

---

[2] Years earlier there were two 680 hp locomotives (off the property since freight service for the CMStP&PRR had ended) equipped to be temporarily fitted with steel plow blades, used only for removing windrows at grade crossings on north side lines, not safe for use near third rail. In the 1978-79 storm it was snow *at the third rail* that was preventing normal operation. The engines were too slow and too heavy to be run over the elevated, so there were a couple of work cars similarly equipped for use at grade crossings on the west side. None of these vehicles would have been useful for the 1978-79 problem.

[3] One observer commented, "It was like driving an auto through salty slush 24" deep and not expecting to lose the transmission or the engine, then taking the car to a garage and expecting it back fully repaired the next morning."

SINGLE TRACK SHUTTLE OPERATIONS CAUSED BY 1979 SNOW EMERGENCY

Route	Route miles	between	and	miles	*Duration of partial operation* starting at	ending at	hours	Reason
Evanston	3.9	Howard	Davis	2.1	1545 Jan 16	2156 Jan 16	6	Blocked by deep snow
		①Howard	Dempster	1.6	1149 Jan 19	0559 Jan 20	18	Running rail tipped at Davis curve
		①Davis	Linden	1.9	1149 Jan 19	0559 Jan 20	18	" " " "
		Howard	Linden	3.9	0559 Jan 20	0957 Jan 21	28	" " " "
		Howard	Davis	2.1	0847 Jan 24	1315 Jan 24	4	Blocked by snow
		②Howard	Linden	3.9	1315 Jan 24	1731 Jan 25	28	" " "
North-South	23.2	Wilson	Howard	3.9	0010 Jan 15	1445 Jan 15	15	Blocked by deep snow
		Wilson	Howard	3.9	1820 Jan 15	1317 Jan 16	19	Ice forming on the third rail
		61st	University	1.3	1728 Jan 15	2129 Jan 15	4	Building adjacent to r-o-w on fire
Ravenswood	9.6	Western	Kimball	1.4	0124 Jan 15	0155 Jan 16	25	Blocked by derailment due to ice
West-South	20.4	Laramie	Harlem	2.5	0714 Jan 4	1300 Jan 4	6	Blocked by derailment at terminal
		Laramie	Harlem	2.5	0418 Jan 15	0505 Jan 17	49	Blocked by snow
		Laramie	Harlem	2.5	0940 Jan 24	1405 Jan 24	5	" " "

Notes:
① *Two separate shuttles were operated end to end and connected by shuttle bus between Dempster and Davis.*
② *Two separate shuttles were operated end to end meeting at Davis.*

rush hours to the irritation of those on the train as well as those unable to get aboard had become a gross *disservice,* not a benefit!

It was painfully obvious that still fewer cars would be serviceable each succeeding day until there was some relief from the weather. It was a dilemma with only one remaining solution: somehow the passenger loading in the peak of the rush period had to be reduced to the level that could be handled by the available rolling stock.

One alternative would have been to suspend all service until enough snow/ice could be cleared and enough cars made serviceable to resume reliable operation of all routes. A procedure of deliberately stopping revenue operation while concentrating on snow removal and equipment conditioning has been followed, in some cities, as recently as early 1993 when a record-breaking storm hit the east coast. Other cities, having a more mild climate, don't try to remove snow in the rare instances when a fall of a few inches suddenly snarls traffic, but simply suspend service and "wait for the guy who put it there to remove it." In the nation's capital, people are sent home as soon as a snow storm is evident and work is resumed only after weather permits. Either of these methods loses revenue from fares but avoids a lot of hard work and some costly equipment damage.

In Chicago's 1978-79 storm, only as the lesser of unattractive alternatives, it was decided to discontinue temporarily making some intermediate stops during rush hours as a means of spreading the severely reduced train capacity over the available network. The stops chosen were those on the inner portions of the Lake and Dan Ryan routes where there was the most alternative paralleling rapid transit and bus service within walking distance. The stations kept open through these brief periods were those in areas that had little or no alternative paralleling service. At no time were any areas denied access to *some* service. Even in the worst of the crisis *all* stations were open during off-peak hours.

The temporary measures proved immediately effective in restoring control and producing a semblance of service. Full time reopening of the affected stations began on the second day and station service got back to normal in time for the afternoon rush of Friday, February 2nd. Every effort was made to communicate the fluid situation to the public through constant updates to the media, some of which stationed personnel in the CTA control center to monitor the situation in real time. Contacts were made with larger employers to update them as to service situations. Staffing was doubled in the transit travel center to provide cur-

Between record breaking snow accumulation and constant overloading, the terrible winter of 1978-79 ultimately caused nearly 100% of the traction motors to fail with the next year. The rebuilding required herculean effort by Skokie Shops mechanics supplemented by work of outside contractors across the country.

Skokie Shop - 2/13/79 - Krambles archive

On a duty for which never intended, a temporarily snow plow-equipped 2000-series car leads a six-car consist fighting snow accumulation on the Lake route. Another unhappy experience came from powdery snow stirred up by train movement entering the self-ventilated motors used on 2000- and 2200-series cars, freezing over air intakes and choking off ventilation.

Austin/Lake - 1-25-79 - Krambles archive

rent status information to telephone callers. Office personnel were assigned on station platforms to help riders similarly.

As can be imagined, during the next weeks there was much criticism, finger pointing and Monday morning quarterbacking of transit performance. Occurring as it did in the midst of a primary campaign for the mayoralty, it became an issue which contributed to the upset victory of one candidate. Much controversy focussed on the part-time closing of the stations, certainly an unhappy choice of strategy that will long be remembered even though the inconvenience it caused was actually relatively little when viewed in the context of the total system status at the moment. Actually the greatest service interruption on the whole system was on the Skokie Swift, where the line was totally blocked for five days starting January 13th and again for three days starting January 24th.

In the 1978-79 Chicago case over a century of previous experience indicated that the extremes of precipitation and low temperature would surely abate after a few days. It seemed

---

TEMPORARY EXPRESS SERVICE ON WEST-SOUTH (Lake-Dan Ryan) ROUTE DUE TO CAR SHORTAGE IN 1979 SNOW EMERGENCY

Tuesday, January 30 - Trains operated express between **69th - Adams** stations and also **Clinton - Austin** leaving 95th, Dan Ryan or Harlem, Lake 0600-1000 hrs and 1430-1830 hrs. At other hours, trains made regularly scheduled stops. Alternative service to Dan Ryan available within one block at North-South line or buses on State Street; alternative service to Lake available on Congress rail or bus lines nearby.

Wednesday, January 31 - Maintenance department reported gain of 10 cars available. Service revised effective PM rush: Trains operated express **69th - 35th - Adams** and **Clinton - Cicero**, from 0600-0930 leaving 95th or Harlem and 1530-1830 leaving the Loop, made regular stops at other times.

Thursday, February 1 - Maintenance department reported another gain of 10 cars. Service revised effective PM rush: Trains operated express **63rd - 35th - Adams** and **Clinton - Pulaski** from 0600-0900 and 1530-1830, making regular stops at other times.

Friday, February 2 - Operated same program as February 1. All stations restored effective 1830 hrs.

The Congress (today's Eisenhower) Expressway was nearing completion when, on July 13, 1957, an unusually severe summer storm struck. Unfortunately, the drainage system of the new highway, with its pioneering median strip rapid transit, neither yet in service, had not been rodded and the automatic pumping equipment was unable to handle the overflow. A torrent poured downgrade for miles eastward into the subway tubes, continuing under the River. It finally came to a balance at the south end of Jackson station under Dearborn Street. At that time the line from Logan Square only extended to LaSalle/Congress. Trains turned through the diamond crossover east of the station, but the inrushing flood waters blocked the area, rising above the lighting fixtures overhead on the platform soffit. The south terminal of service was cut back to Jackson station, and trains from Logan Square were single-tracked from the crossovers at Grand Avenue on both main lines, until cleanup was effected days later.

LaSalle/Congress - April 1951 - G. Krambles

LaSalle/Congress - 7-14-57 - G. Krambles

Near Division/Milwaukee - 4-17-92 - A. Peterson

When the lining of an old former freight river crossing was punctured in a bridge construction error, causing flooding throughout the central area, rail service in the full length of both the State and Dearborn subways was suspended. A bridging service using more than 50 buses was instituted, continuing for several weeks at great expense to CTA and major inconvenience to passengers. This was the north marshalling lineup for shuttle buses serving the gap between the operable portions of the Congress, Douglas and O'Hare rail routes.

reasonable to continue simultaneously and without pause to operate, remove snow and repair damage. However, that clearly proved inadequate, so, for the future it was determined to prepare on the assumption that a storm of this magnitude *would* occur again as climatological conditions follow their secular trend. Although other transit systems had not been hit in the peculiarly hard way that Chicago was, they realized similar vulnerability. With the cooperation of the American Public Transit Association (APTA), a nationwide panel of experts was convened to review the problem and their report, issued later that year, recommended a number of actions.

CTA has aggressively pursued these, adding a unique squeegee-pilot under every driving cab, installing third rail heating in critical locations and obtaining several special purpose snow fighting vehicles (some electrically propelled and others diesel-propelled). This hardware, plus enhanced weather alert capability and vastly enhanced communication capability have put the system in a more comfortable stance for meeting weather emergencies at any time of the year.

Heavy rain can be a problem to the bus system if storm sewer capacity is overloaded and underpasses temporarily flood. On the rail system, Kenton, Douglas, was once a frequent trouble spot and the low area east of Desplaines, Congress, flooded in one storm. Over the years improvements to the municipal drainage infrastructure gradually have mitigated these problems.

One unique emergency for which the system was ill-prepared occurred in April 1992, when a construction contractor working at the Kinzie Street bridge over the north branch of the Chicago River inadvertently drove heavy piling through the roof arch of the former Chicago Tunnel Company line which crosses the river there. Until 1959 this tunnel system operated up to 62 miles of 24" gage, 250 v DC freight railway under the downtown area. The accident quickly flooded basements throughout the Loop and got into both the State and Dearborn subways, causing suspensions of their service, actually for a far greater duration than a century of snowfalls had caused the 'L'. It turns out that Mother Nature couldn't do what a careless human could!

◊ ◊ ◊ ◊

Chapter 12-Employee training and development

When CTA began bringing together independent carriers as a municipal corporation in 1947, among the seemingly insurmountable problems it faced was that there was little emphasis on the most valuable of its resources—the *human* one. Training programs and materials were out-of-date and incomplete. Administrative procedures were documented poorly, if at all, and showed little consideration for the special needs of either the female or minority populations.

A department was quickly identified and staffed to deal with operations and maintenance training. Manuals in the form of **Standard Operating Procedures** were generated to establish, for example, in uniform and time-tested ways, what the bus operator's tasks are and how they are best performed. Tests were simultaneously developed to confirm that the employee comprehends the training and also to assure compliance with it.

With this under way, a **Graduate Training** program was created to recruit new professional and managerial talent from the ranks of recent college graduates, both within and outside of the CTA. It gives them hands-on experience by rotating them within many different departments, leading to placing them in a permanent assignment that best merges their education, skills, experience and interest with CTA's needs. This program has produced most of those who now lead CTA as managers and professionals.

In 1973, at the suggestion of then-Chairman Michael Cafferty, the **Chicago Transit Authority Technical Institute** was begun with the primary purpose of providing one week of hands-on exposure to the real world of public transit for people interested in the field but not necessarily CTA employees. At the same time, CTA employees who served as instructors got to appreciate the importance of their work and to improve their skills.

Cafferty had observed that the newly arising operating, planning and funding agencies were taking on people who, despite strong professional backgrounds, had little detailed understanding of what it took to provide public transit. These folks would be called upon to make important decisions affecting the future and they would benefit by seeing for them-

A complete Boeing-Vertol car is available at the Hawthorne Center to give the hands-on touch to charts, drawings and lectures. The one became available after its mate, 2481, was destroyed in August 1979, when fire totalled the former 61st Street "Whitehouse" shop located on the Jackson Park branch of the North-South route.

Soldier Field parking lot - 6-9-91 - A. Peterson

The International Bus Roadeo of the American Public Transit Association annually compares driving skills of operators throughout the industry. Each candidate's technique is scored in driving a course laid out to represent the typical kinds of obstacles and problems encountered driving a bus in real life. Each bus system participating sends its winning driver to attend (with spouse) the annual meeting of APTA to compete against winners from the other properties for the nation's top award.

selves *"CTA's successes, but also our warts and pimples."*

So CTATI was designed to acquaint students with the hardware, people, problems, strengths, weaknesses, needs and possibilities of the transit system of an ever-changing American metropolis. CTATI has graduated over 1,000 students, with participation from all over the United States, Canada and even a few from overseas. Also, a visiting transit professional from India and another from Sweden each spent months at CTA as interns studying its methods.

A *Management Institute* program was implemented to further develop skills. This was followed by a *Management Development* program partly funded by the UMTA (today's Federal Transit Administration).

By 1974, beginning a more aggressive approach related to the female and minority segment of its work force, a **Human Relations** department was formed to act as ombudsman in issues of alleged discriminatory practices, identifying problems, investigating and recommending solutions. Management review of such issues is thus provided for employees even should they not be represented by a union.

The first females were hired to become bus operators in 1974. In a timely FTA-assisted program, CTA joined the University of Chicago in 1975 to develop a battery of improved standards for the selection of applicants. This facilitated the expanded employment of females as bus or train operating and maintenance workers. Through the application of other training and development programs in place at CTA, many female employees have now worked their way to supervisory and professional jobs.

Uniform guidelines for employment policies such as attendance, conduct, promotion, salary adjustment and discipline are available to employees in an *Administrative Procedure Manual.* The manual is kept up to date by revision as circumstances change.

Once its own equal opportunity standards were in place, CTA joined with other transit properties and the U.S. Department of Transportation to establish for the nation model programs for *Equal Employment Opportunity, Minority/Disadvantaged Business Enterprise* and *Equity in Service Delivery.*

Responding to the plea of one employee, himself a recovering alcoholic, an *Employee Assistance Program* was implemented to help

Driving safety is a key ingredient in training bus and rail operating employees. In a classroom at the Limits Garage Training Center, problems and hazard avoidance in driving a bus are discussed. On the street this is followed up periodically by observation, as illustrated by Instructor Fred Powell and bus Operator Lem Newell. Over the years, through defensive driving measures, accident frequency has steadily declined.

employees who have problems of substance abuse. CTA's EAP, applied proactively as well as reactively in conjunction with established disciplinary practices, helps balance the need to protect public safety with the need to preserve an employee's livelihood.

Selection of Employees

In accordance with CTA's enabling Act, its employees are classified and graded according to job descriptions developed in the first years after takeover from the private owners. These itemize the tasks, skills needed, physical requirements and responsibilities of each job.

Job vacancies are posted to fill vacancies for practically every level of work including professional and managerial. Any employee has the opportunity to compete for advancement whenever he or she feels able to meet the requirements of an opportunity posting. Selection is then made according to criteria established by the using department such as education, experience, skills and work record. In a few cases, final assignment is contingent upon successful completion of special job-related training not otherwise available to the candidate.

Through rationalization policies described in Chapters 4 and 5 and loss of traffic, the work force of CTA was reduced from 24,000 by half between 1947 and 1960. Even so, by the 1950s, natural attrition, longer vacations and competitive employment opportunities made it desirable to improve the process through which new employees were recruited. While taking into account that in some periods there is unusually high or low need for additional staff, new methods were developed to assure proper qualification and equal opportunity for all applicants.

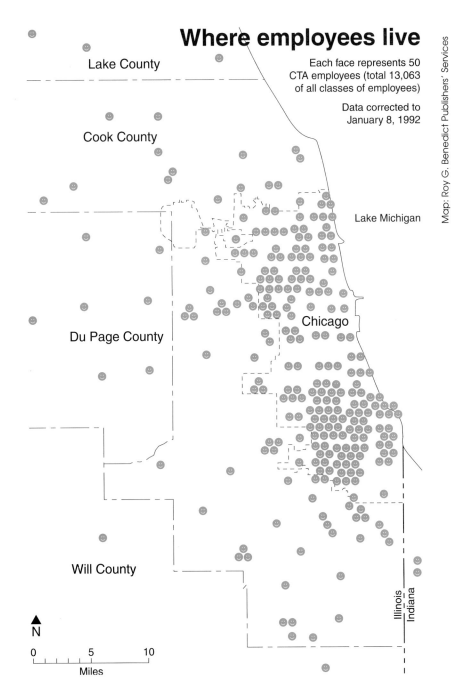

Where employees live

Each face represents 50 CTA employees (total 13,063 of all classes of employees)

Data corrected to January 8, 1992

Map: Roy G. Benedict Publishers' Services

Over the years longer vacation allowances were granted and a summer program of full-time-temporary employment was instituted to increase the number of *summer* vacation slots for bus operators. This attracted college students, high school teachers and others who utilize otherwise free time between semesters as CTA drivers to augment their annual earnings. Once trained, they become expert drivers and many come back to CTA summer after summer. The FTT program benefits not only the FTTs and the regular operators, but it lowers costs as compared to alternative practices of increasing the regular extra board or working overtime on days off.

It is the nature of the transit business that there is a requirement for more service to meet rush hour traffic peaks. In the past these trippers were commonly worked by coupling a rather short morning assignment with another one in the afternoon, spreading the duty of one driver over more than the normal eight hours. In the last few years, a part time classification has been instituted in which a limited number of operators have been employed to fill many of these trippers, making one short morning or afternoon trip on a day.

Training

The transit industry is constantly changing—almost daily something about the routes, equipment, operating conditions and services is modified. The assignment of drivers or trainmen typically takes them to points scattered over CTA's 225-square-mile service area and they often work alone or in pairs under general (not close or direct) supervision.

Whether they drive buses or trains or work at fixed locations, transit employees have complex tools and equipment to operate. Passenger, employee and general public safety as well as accurate, reliable operation requires a training program that includes accurate, understandable training materials. It also demands knowledgeable, perceptive instructors and methods of confirming that employees comprehend and comply with the training given them.

A department dedicated to training was created in 1950. Its first programs for newly-hired operator training were implemented in June, 1952. During those early years, safety emerged as a prime topic of concern. Documents containing standard operating procedures were prepared to describe precisely how to drive a bus or train safely, smoothly and uniformly, a program that has been a major contributor to an exceptionally good safety record.

Over the years training and safety activities have been reorganized from time to time to take on new responsibilities and to enhance capability. Immediate need for its services arose from the new equipment technologies being adopted, for example: trolley buses, propane buses, diesel buses, PCC streetcars, all-electric rapid transit trains, cab signals with automatic speed control, exact change fare collection on buses, the innovative "A and B" skip-stop rail service, new garages replacing old car barns, and the many route changes and extensions.

Not to be forgotten are a series of programs oriented toward understanding and improving of management methods, dealing with the public, supervising others, etc. Among these implemented in the 1970s and 1980s were:

√ - Office employee orientation
√ - Management Institute
√ - Management Education Program
√ - FTA State of the Art Car test program
√ - Public Safety Contest
√ - Standard Operating Procedures program
√ - Garage Superintendent's Manual
√ - Terminal Superintendent's Manual
√ - Bus Operator Security program
√ - Ticket Agent Security program
√ - Wheelchair passenger program
√ - CTA Technical Institute

A significant undertaking in the middle 1970s was the preparation of totally new rule books, the existing old one having become obsolete through changes in transit technology, operating practices and methods of supervision. Also, it badly needed clarification.

Actually three manuals resulted: a **General Rule Book** (governing all employees), a **Bus System Rule Book,** and a **Rail System Rule Book.** While there are similarities in format to books of the past and the sense of the rules is only incrementally changed, the style of wording now conforms to current vernacular rather than the former legalistic tone, and each rule now states up front which employee is responsible for its execution. The style and format of CTA's new rule books has been adopted by several recent "New Start" rail properties.

The training materials and programs just described are administered to the employees concerned by instructors, who

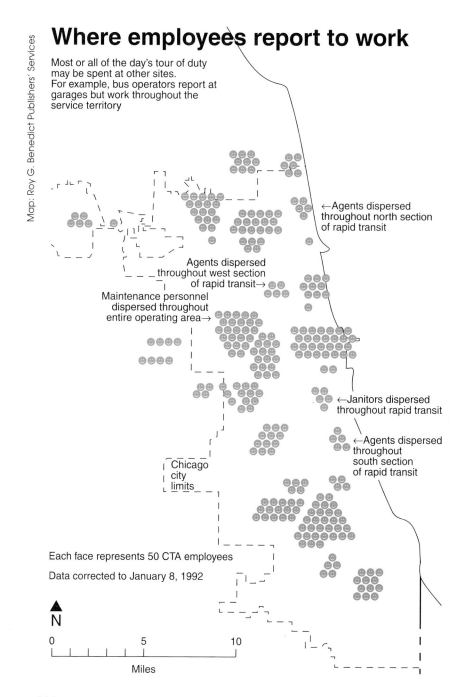

San Diego Trolley at El Cajon CA - 12-17-89 - A. Peterson

In the late 1970s CTA's basic rule books were modernized and updated. Simultaneously, comprehensive documentation of standard operating procedures began, presented in today's vernacular and graphics. Both are components of the effort to reinforce comprehension of and compliance with rules and optimal standard procedures assuring high standards of safety and accuracy of service delivery. CTA's work in these important areas became a model for newly developed light rail systems such as those in Buffalo and San Diego.

certify those who have successfully completed each training element. In the case of certain employees, such as operators, final qualification is dependent on attendance at instruction sessions, satisfactory completion of oral and written tests, practice operation and follow-up riding checks after the employee begins solo driving in revenue service. CTA's certification is accepted by the State of Illinois for the license required of a bus operator. CTA requires recertification periodically. Retraining may be required after an accident, an observed rule violation or a complaint.

The maintenance departments have training and instruction programs tailored to their work. In 1973, part of the old Lawndale Garage was adapted to become a *Training and Behavioral Educational Center* with classrooms and facilities for trainees to learn how to tear down, assemble and trouble-shoot equipment. A replacement, the *Hawthorne Training Center,* now offers air-conditioned offices, 10 classrooms, a lunchroom, lockers, audio-visual equipment and 40,000 square feet of space for hands-on training with components and typical vehicles. Programs supplied by equipment makers are supplemented when necessary by training materials developed by in-house specialists. For the convenience of trainees, classes are offered on day, evening or midnight shifts. The work of the center is supplemented as appropriate by instructors working at the various garages and shops.

Management Development

As far back as the 1920s, Chicago Surface Lines and Chicago Rapid Transit actively recruited engineers from midwest universities and put them on rotational training in various departments as preparation for a transit career. In depression years these programs faded, but when CTA took over it immediately perceived the need to bring the anemic management cadre it inherited up to strength for the job ahead. In keeping with its new responsibilities and the changing times, professionals from more disciplines than engineering were sought. Rotational experiences were changed to reduce the time on the buses and in the shops and substitute instead aspects related to business administration, marketing and accounting.

The field of applicants was opened up to accept people already employed by CTA seeking to upgrade their skills for advancement. Tuition aid is now offered for:

√ - Managerial seminars
√ - Transit industry seminars

Effective training is the key to good maintenance. It is not only essential to the students' comprehension and compliance of the tasks to be performed, but it also facilitates acquisition of the skills needed for a person to progress from vehicle cleaner to electronics technician. Classroom lecture, teardown and reassembly of componentry, attendance at off-site seminars and practice operation (on-the-job training) are all parts of the process. At the Hawthorne Training Center, opened in 1989, Instructor Steven Fronczak teaches Repairman Ronney Hunter, Jr. electrical details of a bus, while Instructor Andrew J. Gasior demonstrates a grade crossing gate mechanism to signal maintainer trainees. Note the ever-present safety training evidenced by the bump or hard hats and, on outside work, high-visibility safety vest.

Both photos: Hawthorne Training Center - 3-28-91 - A. Peterson

√ - Graduate-level business programs
√ - Undergraduate study in universities
√ - Undergraduate study in 2-year colleges
√ - Study in trade and technical schools

Long term employees are characteristic of CTA. Many spend an entire working career of 30, 40 or more years there. Fathers and sons, mothers and daughters abound in the organization and a few families have had three or four generations serving the Chicago transit rider.

Altogether CTA in 1991 has about 13,000 employees of whom 12,000 are directly involved in operating its services or in building and maintaining its plant and equipment. The remainder fulfill the financial, administrative or managerial needs of the system, all of which have increased substantially in recent years due to the expansion of the long-deferred capital improvement program and to tasks associated with the grant processes.

Except for managers, professionals and a few other exempt staff, CTA's employees are represented by the Amalgamated Transit Union or one of seventeen craft unions as their bargaining agents. Compensation meets prevailing wages and benefits, although it is not the highest in the industry. The comfort of a reliable income is a significant benefit of CTA employment; CTA has never been late in meeting a payday.

◊ ◊ ◊ ◊

Oak Ridge/Mt. Carmel - Ca 1909 - by F. E. Borchert, E. Frank collection

Except for occasional rental of its equipment to the Aurora Elgin & Chicago Railway for main line charters, the farthest west reach of 'L' service was once to Oak Ridge and Mt. Carmel cemeteries on the trolley-powered Cook County & Southern branch. High station platforms were provided to accommodate "Met" cars, a few of which, like 800, were fitted with trolley poles. In 1925 regular service was given with an interurban car shuttling from Bellwood 23 round trips a day (26 on Sunday). Oddly, outbound trips all day long ran as train 401; all inbound as 402. On busy summer Sundays extra trains of 'L' equipment operated through to/from Wells, then 5th Avenue, terminal. Funeral charters also used 'L' cars or AE&C 109, which had propulsion/control to run with "Met" cars.

Chapter 13 - Aurora-Elgin, North Shore and others

No history of the *Chicago Transit Authority* could be complete without recording the important impact on its development of the other urban, suburban and interurban electric railways and bus lines in the Chicago area.

The concept of a regional integrated transportation system surfaced about as soon as the first street railway began service. In an earlier chapter it was described briefly how numerous independent companies came together to form operational networks some of which merged to form unified corporate identities. It made business sense to do this to gain the efficiencies of scale as well as to coordinate planning and scheduling. Public pressures to have "one city, one fare" favored such integration—up to the point where political differences between city and suburbs drew a line favoring separate ownership and control, yet essentially agreeing to a certain level of coordination.

Interline operation between connecting streetcar lines reaching to the north, south and west outside the city limits of Chicago began as soon as the lines were electrified.[1]

Similar forces were at play in the intercity transportation field. For the first quarter of the twentieth century hardly any rural roads were little more than mud or gravel. By upgrading streetcar technology for higher speed and for moving freight, the electric *interurban* railway became an attractive way to link towns, cities and the countryside between them.

Chicago-Kankakee

On July 5, 1899 the *Chicago Electric Traction Company,* a local line then operating battery-powered cars from the South Side 'L' station at 63rd Street and South Park Avenue (King Drive) in Chicago as far as suburban Blue Island, opened a five-mile extension to Harvey. The line later changed to conventional trolley operation and extended to the south, ultimately reaching Kankakee in 1908. In 1912, the parts within Chicago (on Halsted and Vincennes) were taken over by *Chicago City Railway.* The tracks below Chicago's city limits became the

---

[1] Before the turn of the century, there was also through streetcar service to northern and western suburbs and even interstate to Hammond and East Chicago, Indiana.

Halsted, Englewood, 'L' station served as terminal for Chicago & Interurban Traction cars and for trippers on CSL's long Halsted route, then carrying through route 13 dash signs. On system maps of the "L" this location tantalizingly carried, in red, the note *"Electric Cars to Kankakee"*. Up on the 'L' is a train of "Big Sprague" cars en route downtown. Until regional shopping centers took over, the most important shopping center in Chicago after the Loop itself was here at 63rd/Halsted.

63rd Place/Union - Circa 1923 - Krambles archive

Chicago & Interurban Traction Company and its cars were rerouted over CCRy to a street level terminal at the Halsted station on the Englewood branch of the 'L'.

That some engineering work was provided by CCRy (and CSL) is evidenced by drawings of C&IT cars and track details from the period 1912-1923. In 1916 design assistance was provided for passenger cars for the ***Hammond Whiting & East Chicago Railway,*** used in through service to the South Park 'L' station. For C&IT one passenger car was designed, which also got help with a home-built freight motor. For its entire life, C&IT continued using its original shop at 88th and Vincennes, which, being well inside Chicago, had been transferred to CCRy ownership in the 1912 deal.

Although Chicago street cars shared tracks with C&IT interurbans, the only regular through service was that of the special hospital car owned by Cook County and designed by CSL in another example of cooperation with outside agencies. This car made periodic trips to move patients between side tracks extended into state hospitals in Chicago and Kankakee.

Gradually, the coming of hard roads, the grade separation of the paralleling ***Illinois Central Railroad*** and its electrification weakened the C&IT. In an attempt at revival at least enough to recoup on a defaulted power bill, it was managed for its last few years by executives (including as president, Britton I. Budd) loaned from the staff of the rapid transit company. During this period two 'L' cars, remodeled as package freight cars were even loaned to C&IT. All to no avail; the weak interurban was abandoned on April 23, 1927.

In the far southwest of Chicago, a joint terminal was provided about 1901 at Archer and Cicero (then the city limits) for the interchange of passengers with the ***Chicago & Joliet Electric Railway.*** When that independent company abandoned rail service into Chicago in 1934, CSL installed a shuttle bus to serve the 3-mile portion of Archer Avenue that by then had been annexed into Chicago. Today this area is served by CTA bus routes 61, 62 and 99. During 1993 it is to also add the new CTA Midway rail line.

---

[2]) In its early years, there were even trips to the Elgin hospital on a roundabout route via Blue Island, Chicago Heights, Joliet, and Aurora involving *five* connecting companies.

There was an unusual juxtaposition of track assignments east of Marshfield tower on the old Metropolitan main line. A Chicago Aurora & Elgin train is shown crossing from the segment where two outbound and two inbound tracks briefly became two paralleling double-track railroads. Then (behind the camera, cutting three slices through the roof of Dreamland dance hall!) they split into a Y-configuration of three double-track lines. No less than nine trains are in this view from a pedestrian bridge connecting the rapid transit platforms to an outbound CA&E platform, also behind the camera. Note the close following distances between trains in pre-ATC-signalling days.

Chicago-Aurora-Elgin

On August 15, 1902, the *Aurora Elgin & Chicago Railway* opened service between Aurora and Laramie station on the west side of Chicago. Here it made across-the-platform connections with the Garfield Park line of the *Metropolitan West Side Elevated Railway (Met),* which extended west from Cicero to the same point two days later.

The Aurora-Elgin line was rapidly expanded with branches to Batavia and to Elgin, and a few years later, to Geneva and St. Charles. There were track connections with other interurbans beyond, but except for a rare special move, there was no through passenger service.

In a reciprocal deal with the Met 'L' in 1905, Aurora-Elgin trains were extended over the Garfield Park line to reach a downtown terminal at Fifth Avenue (later renamed Wells Street) and, at the same time, Met trains were extended west over AE&C tracks to Desplaines Avenue. This marked the first interline operation on Chicago's rapid transit lines. The next year, special funeral service first offered by Met to cemeteries near the Desplaines station was also made available to other cemeteries on the short AE&C Mt. Carmel branch.

A final expansion of interline activity with this interurban, which had been reorganized as the *Chicago Aurora & Elgin Railroad* in 1922, came in 1926, when a new branch was completed by CA&E to Westchester from Bellwood. Service on this was given from opening day by Garfield Park trains extended from Desplaines station over CA&E via Maywood. A final addition in 1930 of about one mile to 22nd & Mannheim station was similarly operated exclusively with rapid transit trains.

After takeover of the rapid transit lines by CTA, failure of these suburban extensions to develop as originally anticipated led in 1951 to substitution of buses for trains on the entire Westchester line west of Desplaines station. By 1953, declining ridership, exacerbated by completion of a paralleling expressway, resulted in CA&E shortening its line to an eastern terminal at Desplaines station and CTA taking over all rail service operating responsibility on the former CA&E segment between Laramie and Desplaines stations. By 1957, CA&E threw in the towel, ending passenger service at noon, July 3, 1957, leaving the people it had carried that morning to fend for themselves going home.

North Shore Line's crack Milwaukee-Chicago Electroliner enters the Loop. As a tenant of the 'L', this connecting interurban railway incurred costs for such things as wear and tear on infrastructure, power, switching, and ticket selling that by 1960 totalled $564,000 per year. Yet Chicago was its main market and it had no alternative to the entry over CTA tracks. The convenient distribution through the central business district over the 'L' with its connecting links throughout the city and suburbs could be matched by no other rail carrier.

#18 Tower, Lake/Wells - 4-14-57 - G. Krambles

Chicago-Milwaukee

On July 16, 1899, the ***Chicago & Milwaukee Electric Railway,*** which started operations with streetcars in Waukegan in 1895 and had been extending southward, began service to downtown Evanston. During the next few busy years, it kept building northward from Waukegan. It also was reorganized successively as the ***Chicago & Milwaukee Electric Railroad*** and then, the ***Chicago North Shore & Milwaukee Railroad.***

By 1908 its service extended from Evanston through Waukegan, Kenosha and Racine to Milwaukee. Parlor and dining service was introduced. By the time the service reached Milwaukee, its passengers could also reach points in Chicago, but had to change to 'L' trains of the Evanston line. Special connecting express 'L' trains were introduced in 1918 after the interurban came under control of the Insull-dominated utility empire in which the 'L' system was a sister company.

Finally, in 1919 after trackage rights and a few essential physical changes were arranged CNS&M trains began to run through over the 'L' to reach a terminal at the Roosevelt and Wabash station. In 1922 some service was extended to the station at Dorchester and 63rd, giving CNS&M coverage in Chicago almost the full north-south length of the 'L'.

In those days, CNS&M's principal market was its hourly high speed interurban service on the 90-mile Chicago-Milwaukee run, but its original Shore Line Route was entirely a surface railway running through a chain of contiguous suburban residential communities along the shore of Lake Michigan. Seeking a high speed by-pass route, a new right-of-way was acquired jointly with the ***Public Service Company of Northern Illinois*** (another sister in the Insull empire), paralleling the original route but about a mile or two west of it in then undeveloped farm land. It was connected to the old route by a five-mile east-west link, completed in March of 1925, extending between Howard station and Dempster Street in Niles Center (today's Skokie). In a reciprocal interline arrangement similar to that with the Aurora-Elgin, shuttle 'L' service between Howard and Dempster stations was inaugurated immediately.

In June of 1926 the remainder of the twenty-mile bypass route (built in only one year) was completed, just in time for the 28th International Eucharistic Congress. A major event of the

Greenleaf Avenue, Wilmette - June 1944 - Krambles archive

During World War II, actually 1942 - 1946, rapid transit equipment made countless trips over the Chicago North Shore & Milwaukee interurban, mostly off-peak and weekends, to serve workers building the camps and boot camp trainees on liberty leave at Great Lakes Naval Training Station. There was a high-level loading platform there. Initially wooden cars were used, typically in six-car trains with one or two trailer coaches in each. Toward the end, only steel cars were used. From December 18-23, 1944, sailors helped move Christmas rush mail at the Chicago Post Office, commuting daily by train of rented wooden 'L' cars. Two trains were used: one had five motor cars and three trailers from Lake Street. A four-car train of Met 2900-series was also used, but gave trolley trouble on the North Shore and was returned after only one day's use.

Catholic Church, it brought at least 600,000 delegates to Chicago. Culminating exercises were held on June 24th at St. Mary's-of-the-Lake, a tiny village located forty miles north of Chicago on CNS&M's Mundelein branch. The steam railroads and CNS&M each planned to carry assigned shares of the anticipated traffic: the **Chicago & North Western Railway** to Lake Bluff, the **Chicago Milwaukee & St. Paul Railway** to Libertyville and the **Minneapolis St. Paul & Sault Ste. Marie Railway** (Soo Line) to Mundelein. All available CNS&M equipment provided service on the old main line to Lake Bluff from points north to Milwaukee and south to Evanston. The final link was provided by shuttle trains that connected Mundelein, Libertyville and Lake Bluff to St. Mary's.

Two 400-ft platforms were built at Lake Bluff for the interchange passengers. A temporary terminal with tracks and platforms to serve eight 'L' trains (with 52 cars) at once was built at St. Mary's. Even hospital and restaurant facilities were included. All for this just one day's productive activity!

According to an **Electric Railway Journal** account at the time :

"In nearly ten hours of uninterrupted loading, the entire throng of 225,000 who had arrived on 372 Rapid Transit and North Shore trains, and an additional 50,000 who forsook motor cars and other means of transportation by which they had come to return by the more convenient electric lines were hauled away from the Mundelein [St. Mary's-of-the-Lake] terminal. The three steam roads, carrying a joint load of only about 75% of the original estimate, were able to handle only 85,000 passengers, or less than one-third of the traffic over the Rapid Transit and North Shore Line. Counting each trip made by a Rapid Transit car, and each one made at least three round trips that day, it was estimated that 15 miles of electric cars were employed in carrying the great mass of pilgrims from and to Chicago.[3] The last train, heavily loaded with company employees and policemen, many of who had been on duty for upward of 24 hours without relief, left the Mundelein terminal

---

3) CRT assigned 932 cars, more than half its fleet, to the special Eucharistic Congress services.

for Chicago at 11:30 p.m....

"By 10 o'clock in the morning—six hours after the migration began—the North Shore and Rapid Transit had delivered 130,000 at the gates of the seminary. Trains carrying anywhere from 600 to 800 passengers were arriving at the Mundelein terminal, discharging their load and started back to Chicago, frequently in less than a minute's time. Including the shuttle trains from Lake Bluff, a trainload of pilgrims entered the terminal regularly every 40 seconds for eight consecutive hours. As late as 3 o'clock in the afternoon special trains were still leaving Chicago as fast as they could be loaded...

"The return movement was made more difficult by a sudden thunderstorm which struck the seminary grounds just as the procession of the sacred sacrament around the shores of the lake was about to start brought a frenzied crowd of fully 100,000 down on the terminal. The situation was acute...the clamorous mob all but swept aside the walls of the stockade in an effort to reach the trains...

"From the control bridge, B. J. Fallon, vice-president of the Rapid Transit Company, and B. J. Arnold, assistant general manager, surveyed the menacing throngs and by speeding up the movement of trains through the "bottle-neck" entrance to the terminal finally succeeded in loading 40,000 passengers per hour and sending one train for Chicago every 55 seconds."[4]

The coordinated team effort of the two sister companies exemplified by the Skokie Valley experience was also characteristic of relations with the CA&E and, to a lesser extent with the Chicago South Shore & South Bend Railroad after it became part of the "Insull empire" in 1925. Technical help from management and engineering staff resident in CRT helped solve design and operational problems on the "big three" interurban roads even to the the middle 1940s; especially with car acquisitions, power improvements and architectural requirements. During WWII, rapid transit equipment was rented to CNS&M to serve special movements, such as carrying seamen from boot camp at Great Lakes to help the Chicago post office handle Christmas mail, workers building barracks, or weekend visitors to the military posts at Great Lakes and Fort Sheridan. For a while CRT's two locomotives even worked North Shore Line freights by day while continuing to move steam road coal in Chicago by night.

CRT in effect provided a training school for department managers at CA&E, CNS&M and CSS&SB, and all the way up to the level of president or general manager. Cooperation continued into CTA years, although constrained by legal requirements separating the municipal corporation from the private companies.[5]

Freight Service

Reference has been made earlier to the operation of 'L' trains over right-of-way leased (and later purchased) from the **Chicago Milwaukee St. Paul & Pacific Railroad** between a point jus north of Wilson Avenue and Wilmette. When this usage began in 1908, the tracks were at ground level, 'Ls' were powered from overhead trolley and freight was moved by the owner's crews, using steam locomotives. A closer relation between lessor and lessee developed in the 1920s when growth of the residential population made it necessary to grade-separate the line and eliminate the nuisances of steam motive power.

As part of the track elevation project, the elevated company then purchased two electric locomotives and rapid transit crews took over operation between an interchange yard built at Buena station and the various industries on the route.[6] The transit company was **not** a freight common carrier, instead it provided a contract service moving carloads on behalf of the steam road. All costs were reimbursed by, and all revenues from the shippers accrued to, the railroad.

The utility of the freight operation declined as the years passed. The principal commodity handled on the freight branch had always been coal and, as building heating was switched to gas, one by one the industrial sidings were abandoned. The cost of moving a car the mile-and-a-half to the last remaining coal yard was pushing $50 by 1973. Small wonder then that the service quit at the end of April that year.

Freight service was also operated by CNS&M over the 'L'. It reached its peak when the North Shore Line began hauling motor trucks and trailers loaded on flat cars, working out of a loading ramp at Montrose, across the street from the freight station used for already

---

[4] The Eucharistic Congress also produced record loads for Chicago local transportation. Monday, June 22, was the biggest day in the history of Chicago Surface Lines, with practically 5.1 million revenue passengers, and the 'L' deployed 1,800 cars, the most it ever moved at one time.

[5] One specialist, MCERA William C. Janssen, an electrical engineer, actually accumulated service on CRT, CNS&M, CSS&SB, CTA and, even the Northern Indiana Commuter Transportation District after it became responsible for capital improvements for CSS&SB passenger service.

[6] Freight operated in both directions on the west track of the elevation, using catenary trolley.

Beginning in 1974 Chicago's suburban bus systems other than those then operated by CTA came under the jurisdiction of the Regional Transportation Authority. RTA immediately initiated bus replacement. CTA assisted, painting three of its new GMC model 5307 buses in varying livery and even having 86 that might be loaned to RTA of a new 600-bus order for itself delivered in "albino" paint scheme (middle bus). Some of these coaches were actually used in suburban service until RTA could acquire its own GMC 4523 "pumpkin" buses (front).

Summit/Touhy in Park Ridge - 7-28-80 - A. Peterson

existing Merchandise (or *Parcel*) Dispatch service for less-than-carload shipments. Motive power (and caboose) for the innovative piggyback trains were motor PD cars, two in front and one at the rear. For many years newspapers and package express were also moved in PD trains, but, except for what could be carried on passenger cars, these services had pretty well dried up before CTA's time.

Other acquisitions

CTA actively considered expansion of its bus system by integration into it of then privately-owned suburban carriers, many of which by 1947 were in even more severe financial distress than the predecessor Chicago companies. Such studies, begun as early as 1948, led nowhere since to pay for any such purchase CTA would have had to issue more revenue bonds, then its only source of capital money. But, under the terms of the trust agreement, to protect the bonds already outstanding, more bonds could only be issued after first certifying that the resulting net system revenues after all operating expenses would would be at least double the total of all principal, sinking fund and interest charges.

This was an insurmountable barrier; revenues of the independent suburban lines barely covered their operating costs, let alone depreciation. Their driver and mechanic wage rates were a little below CTA's, but even presuming that differential could have been maintained in a combined system, economies of scale from integration into the CTA system and route changes due to unfulfilled service demands appeared more likely to net on the negative side. Another generation would pass and additional sources of funding would have to be developed before CTA would venture again into suburban bus expansion.

In 1957, when CA&E was seeking to stop losses by abandoning its lines, CTA was asked to consider taking over some or all of its service. Based on a staff study, a shuttle service using PCC cars between Desplaines and Lombard or Wheaton was offered, provided the requesting communities would find a way to fund the capital costs and the operating deficit. Similarly, when CNS&M was seeking to end its operations, the CTA board offered to provide a high-performance Chicago-Waukegan rapid transit service if its costs were guaranteed by others.

At the time no funding could be found for

A separate public authority, Nortran, the North Suburban Mass Transit District, successor to the privately-owned United Motor Coach Company, operated north suburban bus lines from 1975. Its first order of new equipment was a lot of 74 - 400-series GMC model 5307 piggy-backed on CTA's 600-bus order of 1976. Here the suburban carrier shares CTA's transit terminal at Howard/Hermitage. In 1990 Nortran was integrated into the Pace system, an unusual example of one public authority absorbing another and of operating responsibilities shifting between private and public bodies.

Howard/Hermitage - 3-25-88 - A. Peterson

implementation of any plan to help preserve CA&E or CNS&M service. Nevertheless, the studies did serve to focus attention on the need for public funding to assist metropolitan transit in becoming a viable alternative to highways which were already showing the congestion and environmental damage they could cause.

CA&E's demise in July, 1957 has already been mentioned; CNS&M folded in January, 1963. By that time, the U.S. Congress had created, in the then-Department of Housing & Home Finance, a small program to provide modest grant assistance to a public transit authority willing to test some innovative concept in, for example, service or marketing. How this led to the Skokie Swift mass transportation demonstration project of 1964-1965 and helped gain nationwide momentum in moving today's federal, state and municipal funding assistance programs for transit is told in Chapter 16.

In November, 1973, with partial reimbursement by the City of Evanston in return for a reduced local fare, CTA (using only its own existing buses) began to operate four routes in that northern suburb to fill in where the Evanston Bus Company, a local private operation, had abruptly abandoned service. Because revenue bonds were drying up as a resource, this turned out to be CTA's last suburban expansion. That task was assumed by the *Regional Transportation Authority*, formed in 1974 with funding resources from taxes. In 1984, RTA assigned the management of suburban bus lines ringing the city not then part of the CTA system to its *Pace* service board. A kind of modern descendant of the Aurora Elgin and North Shore Line services of early CTA years is the Pace through suburban services from certain southern and northern suburbs that reach into the Chicago Loop, sharing streets used by CTA routes. Another variant is the service coordination of CTA route 17 with Pace 303 and 310 into Desplaines Transit Terminal on the Congress rail line.

◊ ◊ ◊ ◊

Part of the Lake Street 'L' was built with columns in the sidewalks, but west of Rockwell they are in the street itself, posing an obstruction to street traffic and a hazard to pedestrians. Back in the early days a man was injured trying to board a Chicago Union Traction car at a place where timber shoring was used to support a repair area, posing an ambulance-chaser's dream of a claim between two competing robber baron traction companies. Boarding and alighting accidents were a long-standing problem in streetcar transit. Buses, coming to the curb, reduced the hazards to passengers from other street traffic. The oncoming advent of low-floor vehicles is expected to further mitigate the problem. Rapid transit, with its flush platforms free of street traffic, is safer yet. *(See photo opposite and page 78.)*

Albany/Lake - 5-31-05 - Krambles archive

Chapter 14 - Safety and security

The importance of both safety and security of CTA operations is paramount. Accident and incident prevention is a major goal into which there is a steady commitment of training, supervision and technology. Performance is measured by continuous monitoring of statistical information and potential new problems are detected. Analysis of such data helps spot hazardous locations, shortcomings of equipment and needs for intensified training. Investigation is used to determine the cause of each accident so that effective counter measures can be planned, designed and implemented. Stubborn problems are identified for special research to develop preventive measures.

A small professional group is dedicated to the statistical and analytical aspects of accident prevention. The results of their labors are utilized in the training and retraining of operating personnel and, when appropriate, in the improvement of vehicles, signals, buildings, structures and all other relevant equipment.

Whether, from the CTA point-of-view, an accident is preventable (employee error) or not (e.g., standing bus struck by an automobile), it is reported and recorded. This ensures full accounting and classification of all accidents. Analysis of even the so-called non-preventable accident may suggest a change or strategy to reduce the likelihood of its recurrence.

The accident record has shown consistent improvement over the 43-year record so far compiled, as shown on the graph, page 117. The 75% decline from approximately 27.5 to less than 6.5 accidents per 100,000 vehicle-miles contrasts with a city-wide vehicle accident record of increase from 56,397 incidents in 1948 to 162,544 in 1990, according to Chicago Police Department statistical records.

The rail system, operating as it does predominantly free of the potential hazards of street operation in mixed traffic, has recorded a decline in its accident rate per 100,000 vehicle-miles from only 2 to less than 0.5 in the same period.

To seek a consistently safer future, instructors and supervisors make frequent in-service performance observations of operating employees. Maintenance supports their effort by providing safe, reliable vehicles, track, structures and signal systems. Backfeed from operating experience is used to modify and enhance engineering specifications for new hardware and facilities, while construction of such is carefully monitored by inspectors who demand compliance with recognized industry best standards.

Yet Chicago's safety record has not been without blemish. Until the middle 1970s, Chicago's rapid transit system was the only one in the world that was predominantly without signals for rear end protection of trains, requiring operation "on sight" (as all street traffic does). Careful compliance with operating rules was depended upon to maintain safe separation of trains. Over the majority of the system track mileage there were either no signals at all or simple installations with color light signals but without devices to apply the brakes automatically in case a restrictive signal was not observed.

The record is understandably replete with instances of rear end collisions due to this situation, but most of them were relatively minor, that is to say, involving equipment damage but little personal injury. However, on November 24, 1936 (under CRT management), there was a severe rear end collision at Granville. A heavy steel train of the Chicago North Shore & Milwaukee Railroad (the interurban railway then operating over the elevated) rammed into an eight-car Evanston train. The rear coach, of wooden construction, was telescoped for about 35 feet of its length and part of it fell to the alley below. Ten passengers were killed and 59 injured. Investigation by regulatory agencies at the time recommended immediate consideration of a block signal system and elimination of wooden cars, but lack of funding resources deferred the accomplishment of both objectives for CTA to accomplish four decades later.

The five-mile long State Street subway, opened in 1943, was equipped with state-of-the-art color-light automatic block signals fitted with track trips which enforced a stop should a signal red be passed. This type of equipment was fitted to the Milwaukee-Dearborn-Congress subway

Much of the Lake Street bus route operates between 'L' columns located between the driving lanes, and cannot make service stops at the curb. While 102" wide buses can safely pass each other, extra clearance has been prudently provided through the assignment of 96" wide buses to route 16, the only line on which narrow equipment is regularly used. There have now been two 22-year stints between reequipping the line. In 1969 3700s replaced 1948 GMCs, which gave way in 1991 to 96" wide TMC RTSs.

Lake/Kenton - 1970 - CTA

Looking down on the Lake/Wabash accident of 1727 hrs., February 4, 1977 from the 13th floor downtown offices of the Central Electric Railfans' Association. By dint of a super cooperative effort of several agencies, the site was completely cleared of all wreckage and debris and operation in the area was normal by 0630 next morning.

when it opened in 1951–1958, as well as to a few other critical locations. However, the majority of the system remained unsignalized, or at best, without an automatic train stop device due to lack of capital for the necessary investment for a more secure system.

Development of improvements in signal technology, particularly the audio frequency track circuit, reduced the cost of rapid transit signalling dramatically. Beginning in 1964, this type of equipment, providing continuous cab signal protection with speed limiting control, was installed on the Lake route, where a new fleet of Pullman-built high performance cars was introduced. By 1976, with help from the capital funding programs of the state and FTA (then UMTA), CTA was able to extend the cab signal system to essentially all of the system not then equipped with wayside signals with automatic train stops. Chicago's rail system advanced from having inadequate signal protection to having a most modern continuous cab signal system with overspeed protection. Operating personnel and the public felt a new degree of security.

Unfortunately, on February 4, 1977, the difficulty of providing *perfect* protection became painfully clear, even given the most sophisticated signal system then available. An accident occurred on the sharp left-hand curve at Wabash/Lake, just north of the Randolph & Wabash station. A motorman who had made a normal stop at that station, failed to absorb the fact that, as a consequence of a small delay there was another standing train just ahead, even though this information was being displayed to him by signal indication and functioning of the automatic train stop and speed limiting features of the new signal system.

The unusual coincidental condition of a normal station stop against a restrictive signal due to a train ahead somehow escaped his attention and upon completion of the station work he proceeded (under signal control at a speed less than 15 mph) into collision. Worse yet, following the low speed bump against the standing train, he apparently reapplied power, pushing his 8-car train with 2,500 kw of power against the immovable train. All of this in the middle of a 100 ft. radius curve managed to lift three cars sufficiently to clear the rails and push them outward, falling to the street below with a fourth car hanging precariously by its coupler to the four cars still on the track above. This accident caused eleven fatalities and more than 180 injuries.

The experience, an unusual situation where all the technical equipment functioned exactly as intended and yet a tragedy occurred, emphasized the importance of training, reinforced by supervision and regular means of confirming the comprehension and compliance of personnel with standard operating procedures. It also demonstrated that a new safety device may very well introduce a new hazard. The objective of the safety management in transit is to analyze potential hazards of proposed corrective measures and then to trade off the risks against the benefits to identify the safest overall option, at the same time taking reasonable steps to insure against any known risk. In the case cited, rules and driver training were reinforced and, just in case, additional steel girders to prevent a derailed train from falling off the structure were installed, all to make recurrence of this highly unlikely accident impossible.

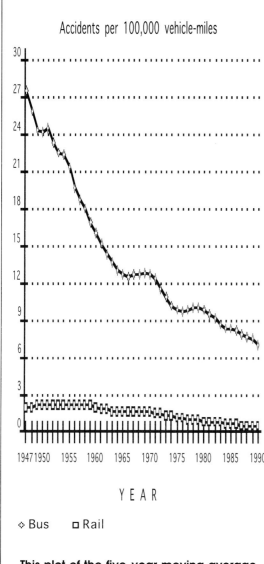

This plot of the five-year moving average of bus and rail accidents shows the remarkable decline accomplished through a vigorous, ongoing safety program involving engineering, maintenance and operation.

Despite these regrettable but rare lapses in rail system performance, its safety record is one of remarkable excellence, as already pointed out in the statistical data. While Chicago's rail system, again unusually among the world's rapid transit systems, has a number of grade crossings with surface streets, it is otherwise free from the interferences of road and pedestrian traffic which plague the bus operation with sudden stops, fender benders, rear end bumps and side swipes.

Unlike buses which must accept road conditions as they are, trains operate predominantly on smooth, straight track, turning comparatively gently and with track transition and superelevation to cushion the change. Potholes don't exist on the rails. Ice and snow seldom interfere with the trains and can't cause skidding to the side. Passengers board and alight from platforms nearly level with car floors and don't have to contend with the high or broken curbs one finds on some streets. The safety department has plenty to think about in the bus operation, but the statistics show that the CTA team effort has been exceptionally effective in improving the result on both bus and rail.

3-28-91 - A. Peterson

Passenger and employee security is enhanced by municipal and private police, some of whom work undercover in plain clothes. Here Transit Detail Officer Charles Klug makes routine interview for possible problems with Operator Jennifer Phinger, one of many such checks made daily. Additionally, the bus radio, supported by a network of locating signposts, comprises a monitoring technology that enables a driver perceiving a security problem to summon police assistance without saying a word.

Chapter 15 - Trains in expressways and trains to the planes

One of the planning innovations introduced to the transit industry at CTA was the concept of integrating a rapid transit line with a major expressway. There are now three such in service here. Electric railway tracks had been laid in the center of streets from the earliest days, with or without pavement between the rails, but relocating an elevated railway into the middle of a modern expressway was something new!

West Side Subway

An excellent summary of the historic West Side Subway project was provided in a brochure issued by the City of Chicago to commemorate its opening for service on June 22, 1958. Excerpting from that brochure:

"Once again Chicago pioneers in a transportation development that is certain to have great influence upon future urban area transportation patterns...

"The new West Side Subway is the first significant project providing rail rapid transit in the grade-separated right-of-way of a multi-lane automobile expressway...[an idea that] greatly increases passenger carrying capacity for comparatively little additional cost...

"The use of the median strip has made possible construction cost distribution of one-fifth for transit facilities to four-fifths for expressway facilities...The West Side Subway project has attracted world-wide attention...[Its] construction...and connection with the Milwaukee-Dearborn-Congress subway was financed by the City of Chicago. Approximately $2 million was obtained by the sale of revenue bonds being serviced by subway rental paid by Chicago Transit Authority. An additional $25 million came from a general obligation bond issue. Chicago Transit Authority, however, is to reimburse the city for the cost of the fixed transportation equipment, estimated at $12.3 million, as it is doing in the case of the State Street and Milwaukee-Dearborn-Congress subways.

"The right-of-way established for the Congress Expressway [today's Eisenhower expressway, Interstate Route 290.[1]] required removal of substantial sections of the Garfield Park elevated railroad structures to make way for highway construction and the relocation and reconstruction of the entire rapid transit route between the Loop and the west terminal in Forest Park.

"Federal highway matching funds, made available by the State of Illinois, and in turn to the City of Chicago and the County of Cook, contributed importantly to financing the cost of the right-of-way and the increased length of local street over-pass bridges [to straddle the rail median].

"The City of Chicago financed and constructed the portion of Subway between the west bank of the Chicago River and Laramie Avenue (5200 West), as well as the new terminal facilities in Forest Park, which will replace the former yard facilities at Laramie Avenue.[2]

"The County of Cook and the State of Illinois, using highway funds, are constructing the portion between Laramie Avenue and the Forest Park terminal, relocating the Garfield Park track and structures in this section. Meanwhile trains are operating over temporary facilities in this area...

" ...The new subway route extends westward [from the River] to Desplaines Avenue, Forest Park, a distance of approximately 9 1/2 miles.

"For the easternmost three-quarters of a mile of its length, the new subway extension is in twin underground tubes. From Halsted Street to just west of Laramie Avenue, a distance of approximately six miles, it is in the median strip of the depressed right-of-way of the Congress Expressway.

"West of Laramie Avenue the two tracks of the new subway pass through curved twin tubes under the eastbound roadway of the expressway. They will continue westward between the relocated tracks of the Baltimore & Ohio Chicago Terminal Railroad and the eastbound roadway of the expressway to Desplaines Avenue, Forest Park, where they recross below the B&OCT tracks and above the expressway to the terminal, just west of Desplaines Avenue...[3]

"The new West Side Subway extension is a two-track facility, but the median strip is wide enough for future expansion. Between Halsted Street and Kenton Avenue (about four and a quarter miles) two more tracks

---

[1] In Chicago, an *expressway* is a grade separated limited-access arterial highway, elsewhere commonly called a *freeway*.

[2] Note that the City of Chicago financed construction of a yard in the suburb because it was needed to replace the existing one inside Chicago. The land this yard is on was acquired by the County of Cook.

[3] The work was completed in steps ending March 20, 1960.

Improved rapid transit service for northwest Chicago

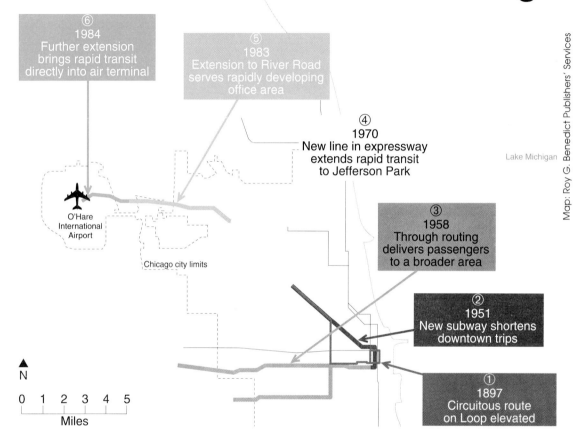

may be added, and between Kenton Avenue and the terminal in Forest Park (about three and three-quarter miles) a third track may be added.[4] *Between Morgan Street and Racine Avenue there is now a third track for switching operations..."*

At Loomis, the west end of Racine station, the two main line tracks widened to accommodate a double track ramp to connect the Douglas branch to the new subway and it was rerouted from its temporary former route to the Loop which had been via the Lake Street elevated.

"Along the new subway extension there are fourteen stations spaced an average of about .7 of a mile apart; ten in Chicago, two in Oak Park, and two in Forest Park...

"...the new West Side Subway is being equipped with an automatic block signal and brake trip system... The tube sections ...including the underpass near Lockwood Avenue, are already equipped with protective signal devices. The remainder of the route will be signalized as the balance of the equipment is delivered..."

Station design on this route is unique, resulting from a compromise between the historic concept of closely spaced stations historically favored by elected aldermen along the route and CTA planners anxious to improve rapid transit performance and system productivity through a combination of widely spaced stations and coordinated local distribution through the bus network.

"Each station platform in the expressway right-of-way is the island type, 600 feet long and canopied throughout its entire length. Supported by structural aluminum columns, the canopy extends beyond the platform edge and over the roofs of the cars..."

Access to each station is at ground level from the middle of a cross street bridge over the expressway. Here there is a building (about 42 ft x 21 ft) containing the fare collection facilities. Connection to the platform below is by a gently sloped ramp. The more important stations are located between two cross bridges separated by about one-quarter mile, and there is an access ramp and at each end. While there are obvious disadvantages to the long ramp concept, the compromise design did avoid additional train stops.[5]

---

[4] To provide capacity for the Chicago Aurora & Elgin Railway, which had trackage rights over the Garfield Park east of Laramie and was the owner west thereof. During design stages of the project CA&E went into abandonment proceedings and the extra tracks were not built.

[5] In 34+ years since opening the line, changing land use and riding patterns have caused total closure of several stations and of the lesser ramp access at others.

Sox-35th/Dan Ryan station -1970 - CTA

Architectural treatment of the ten stations of the Dan Ryan and Kennedy rail extensions opened for service in 1969-1970 was coordinated by the prestigious firm of Skidmore Owings & Merrill to meet City of Chicago (Department of Public Works) and CTA standards. Improved visibility and security, ease of cleaning and more comfortable working conditions for station staff were goals. Employee comfort was a consideration in providing air conditioning and even a compact washroom with a toilet at at one end of agents' booths. Fabricating turnstiles of stainless steel, first used here, has become a worldwide standard in the industry.

With the opening of the reroute from elevated structure to expressway median strip, the name of the rapid transit route was changed from *Garfield* (it had been shortened from *Garfield Park* by CTA) to *Congress.* This and the companion Douglas service, both of which had used the Loop as eastern terminal, were through routed to Logan Square via the Dearborn-Milwaukee subway and ultimately extended through the Kennedy Expressway to O'Hare airport.

Dan Ryan Route

The predicted success of the first median rapid transit line served as a model and Chicago's Kennedy and Dan Ryan (routes **I-90** - **I-94**) and part of the Stevenson (route **I-55**) expressways were also designed to accommodate rail rapid transit facilities. The issuance of $195 million in General Obligation Bonds, approved by Chicago voters in June 1966, allocated $28 million for rapid transit. Utilizing the then-newly authorized programs of the U.S. Department of Transportation provided two-thirds matching funds to clear the way for construction of rapid transit extensions in both the Dan Ryan and Kennedy corridors and for the acquisition of 150 air-conditioned high-speed cars to meet expanded fleet requirements.

Total cost of the Dan Ryan infrastructure was about $38 million and for the 150 new cars cars, a further $19.5 million. The federal grant was announced March 14, 1967, but to fast-track this urgent job, the city's Department of Public Works, the project construction agency, had already begun design. Actual construction was under way by January 1968 and the ten-mile Dan Ryan extension went into service September 28, 1969. Excerpting from the commemorative brochure issued by the city at that time:

"The Dan Ryan line provides a 20.5 mile direct transit route between the south and west sectors of the city. Passengers boarding trains at the 95th Street Terminal can travel to the Loop in 20 minutes. They can transfer to other CTA services along the way, or remain on board the same train to travel west on the Lake Street line to Oak Park and Forest Park, completing the entire trip in 45 minutes.

"The new rapid transit facilities were planned to provide a balanced system coordinated with other existing transportation serving the metropolitan area. Suburban and interurban bus lines will serve the 95th Street terminal of the Dan Ryan Line.

"Nine stations serve the Dan Ryan Line... Wide visibility and a high level of illumination are characteristic features in all areas. Fare collection equipment and turnstiles are of stainless steel and... escalators supplement

stairs for movement between station levels. Stations in the expressway medians are constructed of steel and glass providing maximum visibility from adjacent streets and highways. The boarding platforms are long enough to accommodate 8-car trains... Steel-framed canopies of translucent plastic [extend] beyond the center line of the tracks. Self-service infra-red radiant heaters are located at windbreaks on the platforms.

Off-street bus transfer facilities are provided at the 95th Street terminal and at 69th Street station by means of bus bridges over the expressway traffic lanes. An off-street bus loop is also provided at the Cermak Road station...

"Servicing of Dan Ryan transit cars is provided in a fully equipped carshop with capacity to service 8 cars at one time. The car storage yard south of the terminal has capacity for 130 cars...

"In keeping with the modern design of the entire project, the [150 new] cars are of stainless steel and are equipped with exceptionally large tinted glass windows. Interiors are finished in stainless steel and transverse seats are upholstered in durable plastic. Flush-mounted fluorescent lighting panels and the absence of vertical stanchions convey the impression of well-lighted spaciousness provided in the architecture of the stations...

"Safety in operation is governed by an automatic train control system with audio frequency track circuits delivering signals to the cab. The system provides speed control to match track conditions...

"...Trains will run on continuously welded rails supported by reinforced concrete ties cushioned in stone ballast... [The Kennedy and Dan Ryan lines will be] the first third-rail electric railways in this country to use concrete ties... [They] are used on all main runs. Creosoted timber ties are used at crossovers and other special areas."[6]

Objectives accomplished by the Dan Ryan route include providing a better quality of service to relieve then-existing overloading of the 1892 Jackson Park elevated line and its nearly as old and equally well-worn partner, the Englewood route, as well as to penetrate southward four miles farther toward the city limits.

Ridership on Dan Ryan trains rose promptly to predicted levels and in the generation that has since passed since its opening has continued to grow. Stations on this route are among the most productive in the entire rapid transit system. On a typical weekday more than 50,000 people pass through the 95th station to or from trains, making it CTA's most active. It also indicates the need for further extension of service southward beyond 95th.[7]

However, in the meantime the population served by its mated Lake service remained flat or even declined somewhat. For some years it had been evident that public convenience would be better served and operating efficiency would be gained by interchanging the mating of routes so that the Dan Ryan and Howard services would form one through-routed service and the Jackson Park-Englewood and Lake lines another. This change was an unfeasible cost risk at the time the Dan Ryan was built, was recently implemented with the help of federal and state capital improvement programs that did not exist in the 1960s. The physical changes include costly subway construction, yard expansions, track changes, signalling and other work.

And now to O'Hare airport— Kennedy-O'Hare Extensions

The initial segment of the rapid transit extension in the Kennedy expressway was planned and funded coincidentally with the Dan Ryan application and the two were under design and construction together. However, even though its end-to-end length was less than that of the Dan Ryan, the Kennedy tasks were more time-consuming, requiring a substantial length of new subway including an expressway underpass that had to be built under traffic. Cost of the Kennedy rapid transit extension was about $50 million, not including its portion of the 150 new cars mentioned earlier. It was funded one-third from General Obligation Bond issue and two-thirds from grants of the U.S. Department of Transportation.[8] Service began on February 1, 1970.

The design standards and architectural concepts are generally the same as those employed on the Dan Ryan line with just a few significant differences. These are described best in a few paragraphs excerpted from the brochure which the city issued to commemorate its construction:

"The Kennedy rapid transit line extends 5.2 miles beyond the old Logan Square 'L' Terminal. At Sacramento Avenue, south of Logan Square, the tracks descend from the elevated structure and go into subway, pro-

---

[6] Timber ties were also retrofitted throughout both the Dan Ryan and Kennedy extensions due to problems with the concrete ties.

[7] The city limits, still four miles farther south, remain a hoped-for rapid transit destination.

[8] Opposition to these grants by the Chicago & North Western Railway was withdrawn when CTA agreed not to extend the Kennedy route beyond Jefferson Park. However, years later this agreement was voided.

Engineering and construction of major rail transit facilities is a responsibility of the City of Chicago Department of Transportation (formerly the Department of Public Works). Project manager for the extension to O'Hare International Airport, shown here looking east toward Harlem station, was MCERA Charlie Petzold, who led several other transit jobs. High speed track alignments, wide station spacings and high performance cars make transit speeds competitive with highway travel.

ceeding in a northwesterly direction beneath Milwaukee Avenue, turning north in Kimball Avenue, then crossing under the east-bound and express lanes of the Kennedy Expressway and continuing northwest nearly four miles farther to a new terminal in the expressway at Jefferson Park near Milwaukee Avenue.

"Along the way... three ultramodern stations are provided in the median — at Addison, Irving Park-Pulaski, and Montrose. In the mile and a quarter subway section there are two stations...the Logan Square-Diversey Station, with entrances on Milwaukee at Kedzie and Spaulding near Diversey, and the Belmont-Kimball Station.

"Subway construction...was carried out by the open cut method in three main stages to minimize inconvenience to residents and businesses...

"Coinciding with the start of operations, CTA bus routes serving the northwest section of the city and adjacent suburban communities are being revised for maximum coordination of connecting service at each station... [9]

"The major portion of the new line is designed for speeds up to 70 mph, although the actual maximum operating speed will be 58 mph..."

At Jefferson Park, because the line in the median of the expressway was hemmed in by three traffic lanes on each side, it was not feasible to provide a conventional multi-track yard layout for the cars not needed between rush hours. Nevertheless, sufficient trackage to store 108 cars plus a two-car inspection facility were ingeniously squeezed into the median beyond the station to Foster Avenue. Still, most inspection shop work other than basic trouble-shooting had to be redistributed to maintenance facilities at the Congress and Douglas terminals.

As with the Dan Ryan extension, favorable ridership and operating experience stimulated demand for continuation of the route farther toward the edge of the city and into O'Hare International Airport. This extension, undertaken in the early 1980s, was opened for service to River Road on February 27, 1983 and into its terminal at O'Hare on September 3, 1984.

Evolutionary changes in the details of the O'Hare extension are most noticeable in the architecture of the stations, for each of which a different consultant prepared the design.

---

[9] Among these was the new route #40-O'Harexpress, operating non-stop between Jefferson Park and O'Hare airport.

The four stations are at Harlem, Cumberland, River Road and O'Hare, with an average spacing of about two miles.[10] All of them have island platforms and are provided with elevators to comply with current accessibility standards, as well as escalators and stairways.

At the first two, where the median is depressed, "box-frame" canopy construction is used, in which all supporting columns are on the highway side of the tracks, leaving the platforms totally free of such obstruction. At River Road, the I-90 expressway approaches a five-way cloverleaf of highway viaducts, interconnecting with the North-South tollway (I-294), and the branch into the airport (I-190).

The 12-car Rosemont inspection shop and 260-car capacity Rosemont yard (replacing the paltry temporary facilities at Jefferson Park) were both fitted into previously unused segments between these expressway ramps. Because of the track configuration leading into the yard from the west end of River Road station, the platform there is wedge shaped. It is also covered with box-frame canopy construction.

West of Rosemont yard the route descends into subway for its final approach, passing an 8-car center track for holding a reserve train. It ends in a spectacular three-track terminal station in a cavernous space, containing the two platforms and three tracks totally free of columns. Cleverly designed into the already-existing parking structure centrally located between the several passenger terminals at O'Hare, it represents an example of remarkable engineering and architectural achievement.

At O'Hare terminal, in the "unpaid area" beyond the fare controls there are enclosed passageways leading into the four principal airline terminals, in the trunk route of which there are moving belts to facilitate passage. Now nearing completion is a new International terminal and also a Matra-designed automated people mover which will interconnect the various terminals and parking areas.

As an adjunct to the rapid transit line, there is off-street auto parking at Cumberland and River Road stations, the former in a multi-level structure currently being enlarged to meet growing demand.

The O'Hare project was CTA's first extension into a relatively undeveloped land area in more than a half-century. The results have been encouraging, stimulating a remarkable development of residential, commercial, office, convention center and amusement activities nearby.

It also confirmed the wisdom of direct rapid transit service to a major airport. Those who thought that, at the most, only airport workers would use the service have been surprised to discover the substantial use of the service by airline passengers who have discovered its convenience and dependability. O'Hare station is among the top third of CTA's 143 stations in number of riders served. In a broader view, experience is also proving the merit of wide station spacing as used on the O'Hare extension in optimizing service appeal and operating efficiency.

and to the the other airport—Southwest Transit Project

Currently approaching completion is the $500 million Southwest rapid transit route extending into a quarter of the city that has never before enjoyed rapid transit service. In general, the route parallels the Stevenson expressway (I-55), but unlike the previous projects, it predominantly follows present and former railroad rights-of-way. The new line starts from a connection with the Dan Ryan 'L' structure at about 18th & Federal Streets, then follows Archer, Leavitt, 49th and the Belt Railway of Chicago to a station at about 4600 west on 59th Street, a few hundred feet from the airlines' passenger terminals at Midway airport. Facilities at the rail terminal include a yard and inspection shop.

There are seven stations about a mile apart on the nine-mile extension, and a new station at Roosevelt/Wabash on the existing elevated. The service will circle the Loop clockwise to contact the nine present stations there. Of an order of 256 new cars of the 3200-series already being delivered, about 90 will be assigned to this route when it begins operation, probably late in 1993.

Current planning for a third airport in the southeast corner of Chicago by government at federal, state and city levels has raised speculation as to the future of Midway as an airport, but there is little doubt that the residential, commercial and manufacturing activities of the southwest side will be substantially stimulated by the new rapid transit route and can support it with or without the airport. As to the new *southeast* side airport being advocated by Chicago's Mayor Richard M. Daley, it would certainly involve rail mass transit, probably through extension of the Dan Ryan line via the I-94 expressway, plus enhancement of the Metra Electric-NICTD-South Shore Line corridor via Hegewisch.[11]

Central Area Circulator

The "Circulator" project of the City of Chicago, currently funded through preliminary engineering, would design and construct an eight-mile light rail transit (LRT) system linking the North Western, Union and Randolph Street

---

[10] Due to the unusual way in which lands, including O'Hare airport, were annexed to the city, *all* the stations are *within* Chicago except River Road, which is in the suburb of Rosemont.
[11] NICTD trains also pass the Gary airport and directly feed the Michiana Regional Airport at South Bend.

Metra suburban stations to North Michigan Avenue, Streeterville, Navy Pier and McCormick Place. The system would include east-west routes north and south of the Chicago River as well as north-south links on portions of Michigan Avenue and Columbus Drive.

Selection of consultants for preliminary engineering and program management have been under contract since September 1992. According to timetable, preliminary engineering would be completed by the end of 1993. Following this, the city would secure funding to complete design and construction, with operations expected to begin in 1998.

A fleet of 66 low-floor cars is envisioned. These cars would be capable of a 50-mph top speed, although only a few short private right-of-way sections would exist. Except for the North Riverbank Corridor, the system would operate in dedicated lanes in curb lanes of city streets, most of them one-way with priority traffic signalling at intersections. The average station spacing would be 0.28 miles and the end-to-end operating speed, 8 - 15 mph.

Total construction cost has been estimated at just under $650 million. Operating costs are predicted to be $20 million annually by 2010, with cost-recovery ratio about 50% matching the overall RTA system goal which currently mandates that CTA recover 52%.

Although under its enabling ordinance CTA has the exclusive right to operate local transportation in Chicago, its relation to the Central Area Circulator project is not yet decided.

While not an expressway-related project, the Central Area Circulator, representing a return to operation of rail transit in streets in mixed traffic would be, for Chicago, an interesting (and challenging) renaissance of multiple modal use of surface arteries.

◊ ◊ ◊

Taxiway at O'Hare International Airport - 1984 - CTA

Rapid transit lines built before about 1923 could be expected to stimulate land use causing population growth and new transit ridership. A veritable explosion of automobile ownership plus the collateral effects of the great depression ended that. Single car train with #1044 betrays overdevelopment of the network after the Niles Center branch opened in 1925. Yet rapid transit service on this line, which had been replaced by motorbuses in 1948, was actually revived by the non-stop Skokie Swift in 1964. The buses were retained for local distribution while the trains now operated non-stop.

Oakton/Niles Center - 6-12-30 - Krambles archive

Chapter 16 - The Skokie Swift

Skokie Swift is a rail rapid transit shuttle service that developed as one of the first mass transit projects in the United States involving cooperative sponsorship of the federal government and a transit operator. As a "demonstration" project (in the jargon of 1964) authorized by the National Housing Act of 1961, the service was provided on an *experimental* basis during the two years beginning in April 1964.

The principal objective was to determine the effectiveness and economic feasibility of linking a fast-growing suburban area of medium density (Skokie) to a central city (Chicago). It became possible due to there being an existing stretch of compatible electric railroad owned by the 1963-abandoned Chicago North Shore & Milwaukee Railway, already connected to the CTA system. In addition to having been part of the main line of a 90-mile interurban passenger and freight service, about three miles of this track was used under trackage rights by CTA to provide access to its Skokie shop.

The project would offer a rapid transit shuttle train between Dempster Street, just outside of the business district of Skokie, and the nearest rail transit terminal at Howard Street, the city limits between Chicago and Evanston. There were to be no intermediate stations but service would be coordinated at each terminal with the existing network of rail and bus routes (one of which also provided local distribution in the intermediate area).

Through surveys and studies, the project had the further objective of determining guidelines and criteria that might be useful to public officials, planners, transit operators and others in future consideration of other services of this type.

Funding sponsors of the project were the Village of Skokie, the Chicago Transit Authority and the U.S. Housing and Home Finance Agency. During the term of the project, the HHFA became the Department of Housing and Urban Development. Still later, in July 1968, its transportation programs and activities were transferred to the newly formed Department of Transportation under its Urban Mass Transportation Administration (today's FTA). The Chicago Area Transportation Study and the Northeastern Illinois Planning Commission supported the project by conducting and analyzing the survey and study fact-finding program.

As a preliminary to the project CTA drafted a plan for the operation which set forth the anticipated minimal needs for rehabilitation and improvement. It also drafted a work and study program. Using funds from its depreciation reserve, CTA purchased five miles of electric railway that had been abandoned and without operation for more than a year. Then, aided by a grant from HHFA, the project team rehabilitated this property and modified sufficient rolling stock and other facilities to permit the planned operation.

A "Park'n'Ride" lot with a capacity of 555 cars, to operate with a fee, was constructed at the Dempster Street terminal in Skokie. It included a "Kiss'n'Ride" area (novel at the time) where transit riders are dropped off and picked up by auto. Space was also provided for the loading and unloading of buses, including those of unaffiliated suburban and intercity carriers, serving as feeders to the rail service.

The project plan provided fifty round trips at 15- or 30-minute headways between 0600 and 2200 hrs., Monday-through-Friday, except holidays, with no service on weekends. Four experimental high-speed cars were assigned to protect a schedule that demanded only two cars in the peak.

Patronage on the project's opening day, April 20, 1964, was 3,939 riders and by the end of its third year rose to 7,500. The comparable area had provided only about 1,500 riders per weekday in the last year prior to its January 1963 abandonment as part of the Chicago-Milwaukee interurban railway. Obviously, the planned schedule and fleet assignment plans were scrapped from the first day, when 75 trips were run, and this was augmented several times. Based on this, Saturday service, not originally planned, was offered from the start.

Skokie Swift trains are of driver-only operation. Initially, the trains consisted of single cars, but in 1965, four 18-year-old three-compartment articulated cars were modified and introduced in the project. After twenty years of further service, the "artics" were superseded by married pairs rehabilitated from existing 1959-vintage cars. During the first years of Skokie Swift a maximum operating speed of 70-mph was attained and this yielded a schedule speed of 46-mph over the five-mile non-stop course, making the Swift "the fastest rapid transit train on earth", to translate the words of a German rail magazine of the period.

Difficulties of maintaining reliable performance with the aging infrastructure and the prototype high-performance rolling stock later were offset by reducing the maximum speed to 50-mph, but with its non-stop run over the five-mile course, Skokie Swift maintains a respectable average of more than 35 mph, one of the highest terminal-to-terminal speeds in the industry.

During 1991, in an initiative of the Village of Skokie, the parking lot and bus terminal was augmented by an addition on the north side of

FINANCIAL RESULTS of the SKOKIE SWIFT PROJECT
Project life - April 20, 1964 through April 19, 1966

	Trains	Park'n'ride	Total
Revenues			
Passenger fares	$704,416	$71,486	$775,902
Rental of easement	16,800	--	16,800
Other	4,581	--	4,581
	$725,797	$71,486	$797,283
Expenses			
Operating	$479,384	$44,296	$523,650
Depreciation on cars	56,916	--	56,916
	$536,270	$27,190	$580,566
Net Operating Revenues	$189,527	$27,190	$216,717

Rehabilitation expenditures	
Signals and protection at seven grade crossings	$117,670
Telephones	2,926
Line supervision (including automatic train despatching)	3,430
Train phones	4,630
Catenary and power distribution system	32,127
Skokie substation	10,431
Dempster bumping post	244
Dempster turnout	8,187
Other track and grade crossing work	43,663
Snow melters	1,387
Right-of-way fencing	5,513
Demolish unused Main, Oakton, Kostner, E. Prairie stations	16,974
Construct new (temporary) Dempster station	41,654
Modify cars (pan trolleys, etc.)	51,474
T O T A L (Net project cost after two years)	**$340,332**

Here are some of the CTA people on hand to assure successful operation of Skokie Swift on dedication day. For the inaugural train, all four cars then modified for the service were coupled together. On the outbound VIP trip, when a pan/trolley was inadvertently ripped off by raising it just a little too soon, men from this group quietly got the wounded car to the Skokie shop. Without even taking time to change into their work clothes, they repaired the car and returned to Dempster terminal in less than the 45 minutes taken by the ceremonial speeches! Only employees and railfans were aware that anything had challenged the success of the day.

PKS Kiewit Western contractor's men spread new ballast on Skokie Swift. Rebuilding of at-grade trackage here and on the Ravenswood and Douglas lines in 1991 set new standards for construction and use of modern materials. Indirect track fixation and geotextile fabric under the ballast should reduce long-term maintenance needs and lengthen component life.

Dempster/Skokie Swift - 4-18-64 - CTA

Main/Skokie Swift - 4-21-91 - A. Peterson

Dempster Street. Then, on the CTA's capital grant program, track over the full length of the route was renewed. Unfortunately, this involved suspending the rail service for some months. In other Village and Illinois Department of Transportation initiatives, a new Skokie terminal rail station will replace the temporary 1964 shelter, and there will be improvements to the adjacent bus turnaround. When this is followed by new, air conditioned cars replacing equipment rebuilt from components forty-some years old, the Skokie Swift may be expected again to be a pacemaker among smaller transit operations.

The 1966 conclusions of the project were that a median-density suburban area can be linked effectively with the central city by a high-

Dempster, Skokie Swift - April 1993 - G. Krambles

In mid-1993 an impressive new Dempster station was approaching completion to replace the "temporary" facility that had served there for 29 years. Design and construction supervision was by architects Dubin, Dubin & Moutoussamy, who have done much transit station design in Chicago, including on the Southwest Transit Project and numerous other individual stations. Partner (and MCERA) Arthur D. Dubin, principal for the project, has been a Skokie Swift supporter from inauguration day.

speed rail rapid transit extension on an economically feasible basis provided that public investment provides the right-of-way, equipment and infrastructure. It showed that under certain conditions many stations, long trains and direct through service are neither necessary nor desirable in providing a viable service and that a significant number of riders will switch from their automobiles to transit if the alternative, like Skokie Swift, provides a convenient, reliable and time-saving service. It also clearly showed that a Park'n'Ride facility which offers easy access from expressways and major arterial streets to an outlying rail terminal is highly important to the success of such a service.

Skokie Swift, with its close two-way headways in both rush periods, has opened new job opportunities, a point of special importance to less advantaged inner city residents who cannot afford automobiles to drive to suburban jobs, nor, alternatively, to find housing in the suburbs. Reverse peak riding is now significant.

It showed the advantage in first cost of utilizing existing railroad rights-of-way made available by changes in long-haul freight and passenger travel demands. In times past, planners recommended only buses for loads up to about 5,000 people per maximum hour and thought in terms of 10,000 to 40,000 or more as being needed to support rail transit. But people in less populated areas also seek choices in addition to driving. Of course, they expect alternatives to be attractive, in tune with modern standards of speed, convenience and dependability.

Skokie Swift, with a peak hour load of about 1,500, proved able to support a far higher quality of service than a bus that had to share its streets with general traffic. Whether buses could move the volume seems unimportant if they do not generate it, as was the case before the Swift started. However, the bus in parallel with the rail provides a different service, distributing to local stops en route and providing a supplemental week-end coverage.

Although it remains one of the world's smallest rapid transit routes, Skokie Swift success encouraged the creation of local, state and federal programs to assist transit capital projects elsewhere. It helped set the pattern for the many new-start rail systems, becoming really the first of the modern *light rail* transit lines without ever knowing that name!

In a further development of the 1990s, interest was growing in extension of the Swift, probably a further eight miles north to the Lake-Cook county line, where there is a substantial complex of shopping and corporate offices, and also southward to an interchange with the CTA West-Northwest Route near the Montrose station for crosstown access to the northwest side of Chicago and to the O'Hare International Airport. This local initiative has been incorporated in the year 2010 transportation plan of the Chicago Area Transportation Study, the regional Metropolitan Planning Organization, and it now awaits the necessary capital funding.

◊ ◊ ◊ ◊

Chapter 17 - Ridership, marketing and fares

There is no doubt that nationwide overall ridership on public transit has suffered a long term decline.[1] According to the American Public Transit Association, transit ridership grew steadily from 1900 to 1929, due principally to population growth, technological innovation and investment opportunities captured.[2] Economic depression is cited by APTA for the sharp decline from 1929 to 1939. To this must be added the competitive influence of the introduction of paved roads and the explosion of private automobile ownership, which together practically extinguished the intercity electric railway business even in the face of massive growth in the total number of passenger trips being made.

A new federal law limiting the ability of utility companies to subsidize transit (or vice versa) as had been normal since the very beginning of electric transit, led to decline in access to capital for renewal or expansion.[3] World War II brought motor fuel rationing and an economic boom that revived transit ridership but at the same time wore out much of its physical plant.

Postwar public policy reversed the trend as abruptly, sending riding sinking even more deeply than had the depression. Beginning in the 1950s construction of the interstate highway system, together with cheap gasoline prices, made intercity and suburban auto travel more attractive than ever and this eventually dried up cross-country rail and bus traffic. Important segments of the interstates within the great metropolitan areas lured lots of riders into autos and away from transit. True, the expressways helped bus operation by relieving some basic urban streets used by transit of some of the choking auto and truck traffic that was hampering schedule performance. But arguably the greatest loss for transit was the combined effect of the highway system and public taxing and financing policies which resulted in massive shift of jobs and residences from city to suburbs.

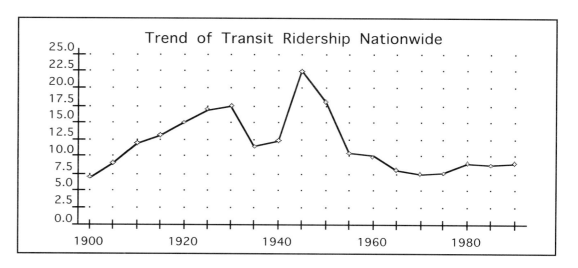

Gradually, unfavorable impacts of some of these public policies on the real costs and qualities of urban life began to be recognized. Undesirable side effects such as worsening environmental pollution and expedited energy depletion became obvious. Problems are still being caused by the uncontrolled shift of whole segments of population and workplaces away from central cities, leaving behind a declining economic base to deal with increasing social costs.

In the 1970s these conditions led to developing a partnership of local, state and federal governments with a willingness to test ways to improve transit and halt its attrition. Unfortunately, under the impact of current deficit spending and depressed business conditions, public funding capacity is shrinking and once again a downturn in ridership seems like to be the consequence.

In spite of the difficulties, CTA early-on organized a specific marketing function to provide public information, press, radio and TV media support, maps, brochures, public speakers for community groups and many other kinds of activity to keep the communities served informed on services, changes and even, when necessary, real-time condition reports when there are emergencies or unusual traffic problems. In the middle 1970s CTA travel telephone information was transferred to the Regional Transportation Authority and expanded to include the suburban trains and buses operated today by Metra and Pace.

---

[1] This decline is difficult to measure precisely because of mergers, changes in fare structure and fare collection methods. For the same reasons, evaluation of the effect of pricing variations, such as monthly passes and discounts for senior and handicapped riders, is imprecise.

[2] According to APTA 1991 Transit Fact Book.

[3] The Public Utility Holding Company Act of 1935, administered by the Securities & Exchange Commission.

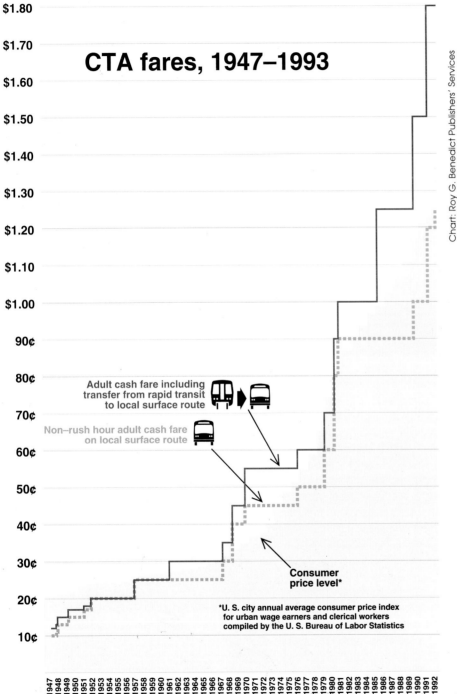

Of course, in a broader sense, marketing includes important ingredients contributed by service planning, operation, maintenance and engineering. Long range strategic planning is also essential. All of these capabilities have been developed and applied by CTA. Every CTA employee or Board Member of necessity is part of the marketing effort, and in a sense, so are all its riders and suppliers!

Fares policies

User charges for transit can be in the form of a flat fare for any length of ride, or a graduated fare that increases with trip length. One other basis that has significant acceptance in the urban transit field is the pass, usually good for unlimited riding over a certain period, such as a week or month.

Basically, Chicago's streetcars, rapid transit trains and city buses have historically been on a flat fare system. The early horse car lines had a simple flat nickel fare, one rate for all riders except for infants, who rode free. There were no transfers; a new fare was due on each vehicle. A trip of as little as a mile cost double fare if one had to change cars en route. When the transit network grew, as a condition of granting franchise ordinances, companies were soon forced to provide transfers between their own lines.

Prior to the Unification Ordinance that was effective February 1, 1914, there were five separate companies authorized to operate in the City of Chicago.[4] Each had its own certificated streets or rights-of-way and charged a 5¢ fare on all its routes, except for certain "through routes" on which there were free transfer arrangements between the streetcar companies. One company charged a 10¢ fare for a ride through onto its lines south of 79th Street.[5]

---

[4] They were: Chicago Railways Company, Chicago City Railway, Calumet & South Chicago Railway Company, Southern Street Railway, and the Chicago Elevated Railways (a voluntary association of the underlying properties).

[5] Calumet & South Chicago Railway Company.

Fare collection methods have varied over the years. For example, on the bus system, in the early 1950s some routes were equipped with fare registers while others had fare boxes. Of Brill-built trolley coaches 9203-9322, those up to 9286 had farebox stands and register brackets, while most others had only the brackets. Morning and afternoon fleet requirements varied on some routes, so fare collection equipment had to be juggled between buses accordingly. Operators were responsible for counting and remitting all fares collected during their runs until 1969, when the exact-fare system became practically universal throughout America. Today, all buses have sophisticated electronic fare boxes that count the money and show the number of passengers by fare paid.

Both: North garage - 1953 - G. M. Andersen

A passenger needing to use certain combinations of routes would have paid at least two fares. Using both streetcar and 'L' on the same trip would have cost two fares until as recently as 1935, when a limited number of free transfer locations between those modes was ordered by the Illinois Commerce Commission.[6]

There has been a recurrent interest in fare systems that would price a ride in proportion to the distance travelled, or at least charge more for a long ride than a short one. There are technical problems that make this difficult to do, but given advances in microprocessor technology and magnetic card reading, they may one day be solved. The cost and time to implement would

---

[6] Universal transfer, making every common stop between streetcar, bus or 'L' a transfer point, was achieved in 1943.

be substantial, but may not be the main constraint. The institutional problems, unfortunately, are less tractable.

For one thing, a century of precedent with flat fare has given a semi-sacred aura to the concept of, *"One city, one fare,"* not actually achieved until October 1, 1952, under CTA administration.[7] Given the forces on transit management of inflation, improved standards of service demanded by the public, improved standards of living demanded by the work force and shrinking market share in a growing trip population, fares, which had remained almost constant for a half-century started to climb.

For a variety of reasons, but chiefly because of massive expansion of investment in automobiles and highway systems, as fares rose riders were lost as transit customers. Short haul riders diminished more rapidly than those taking long trips since buses, and especially rapid transit, still offered attractive time and comfort advantages over driving. When revenue had to be increased to cover rising costs, was there a way to minimize the impact on short haul riders so as to avoid diminishing returns from that group of riders due to a fare increase?

From time to time the CTA Board studies whether zone or distance-based fares should be adopted. The technical problems of administering such schemes on CTA's complex gridiron network of bus and rail lines (some very fast and others painfully slow) have already been mentioned. Getting the rules and procedures understood and complied with by the public and the employees would be another major difficulty. And there can be some argument whether a rider perceives sufficient extra benefit from a long ride compared to a short one to be willing to pay proportionately.[8]

---

[7] The slogan was almost never literally true; variations over the years included: zone fares outside the city, downtown shuttle fares, express service surcharges, age-based and handicapped reduced fares, term passes, etc.

[8] Short trip riders who go through peak load points on long lines may actually be the most costly to the Authority since even though travelling only a short distance, they set the maximum amount of equipment that must be operated.

Modernized "Big Pullman" car 204 advertised the CSL-CRT intercompany transfer about to be permitted at 52 station locations mandated by the Illinois Commerce Commission to improve the integration of public transportation. The concept was expanded in the next three years to more stations, to include CMC, and even, on a trial basis, to 12 stations where streetcar or bus lines of the Chicago & West Towns Railways intersected CRT. To limit stopovers at transfer points, CRT required time-stamping the transfer slip on leaving the 'L' and for this installed hundreds of red topped transfer stamping machines, most of them made by IBM.

Part of the South Shops complex, the transfer print shop daily produces about a million transfers for passengers' use to extend their ride on connecting buses and trains. Since transfers have a cash sale value, security precautions protect production and distribution. Another important specialty function nearby is Central Counting, where each day's cash receipts, received in bulk from the garages are sorted and counted. Transit's daily turnover of silver is the largest single source of change anywhere in the city; retail business in general depends on it being promptly returned to circulation.

Before regulations forbade the use of federally-funded buses in charter service, such traffic generated 1.5 - 2% of the total revenue received from system users. In a record-breaking movement CTA carried 38,400 riders on 800 buses to and from the Marian Year observance. Buses of the North Shore Line, Evanston Bus Company and South Suburban Safeway systems are also in this view.

In view of the political sensitivity involved, changes in fares are customarily made only when additional revenue is an immediate necessity for continued operation, so they can no longer be deferred. Given this constraint, probably the most challenging aspect of introducing distance-based fares would be how to generate the needed new revenue without pricing a long ride ***drastically*** more expensive than a short one, since the vast majority of riders are only riding a few miles. If flat fare policy has cost dearly in short haul traffic, distance-based fares risk losing long haul patrons.

◊ ◊ ◊ ◊

78th/Vincennes - January 1974 - G. Krambles

Soldier Field - 9-8-54 - Krambles archive

Jefferson Park/O'Hare station - 6-9-91 - A. Peterson

Intermodal terminal was designed to be shared by ten CTA bus routes, CTA trains, Metra's C&NW-Northwest commuter line, suburban buses of Pace, and intercity buses of Greyhound. The policy of providing similar facilities for convenient coordination between transit services, initiated for Chicago by CTA, is evident in every rail or bus project in the region.

Chapter 18 - Partnership in the metropolitan network

The three parts of the story of the Chicago Transit Authority are—what came before, today, and the future. In the early chapters of this book there is a capsule view of pre-CTA days and a summary of what's gone on in the 45 years since. In this chapter the whole of the public transportation network of the Chicago region and CTA's relationship to it is examined.

Although *individual private entrepreneurs* vision, investment and perseverance created the concepts and basic technologies of streetcars, buses and trains, there was always a close correlation of interest with the *public sector* which developed and managed land use, an obvious cause-and-effect impact of the decisions made by the owners of the transit systems and the elected representatives of the communities. Transit service supported the development of residential, commercial, educational and administrative land use and vice versa. When good ideas were implemented on either side, the other benefitted. If there were shortcomings on either side, the other suffered.

For most of a century beginning in 1859, problems in the development of local mass transportation brought about decreasingly satisfactory relations between the factions. Some unification was mandated in 1912-1913 on the private owners of streetcars and rapid transit. Private ownership held on until the 1940s, but then the handwriting was on the wall. Public ownership provided a fresh start in 1947 when CTA was created with more freedom to decide fares and service levels, taking over most of the streetcars and buses in Chicago and contiguous suburbs. The new approach worked reasonably well for about a generation, but began to break down in 1970, when system-generated revenues, coming primarily from fares could no longer be stretched to cover operating costs, renewal of worn out equipment, and the investments needed to serve the expanding community and the evolving transportation patterns.

The problem was not confined to CTA but was endemic to the commuter railroads and suburban bus lines throughout the six counties comprising the urbanized area of northeastern Illinois.[1] On December 12, 1973, the **Regional Transportation Authority** was created by the Illinois General Assembly as an operating entity to coordinate and provide a consistent level of financial support to his region. In 1983 the RTA

---

[1] Cook, DuPage, Kane, Lake, McHenry and Will counties.

Act was amended to define three subsidiary agencies (service boards)—in addition to *CTA*, the commuter rail division, *Metra*, and the suburban bus division, *Pace*. RTA itself was reconstituted as a planning, funding and oversight entity over the three service boards.

Funding of the transit system comes primarily from passenger fares, other user charges, federal operating assistance, and sales taxes.[2] The General Assembly also imposed on RTA a requirement that system-wide at least 50% of all operating expenses of the three service boards must be recovered through farebox revenues. In

---

[2] Financial stability was secured by the allocation to RTA (restructured January 1, 1990) of a percentage of general sales taxes in Cook County, and a smaller portion in the collar counties. Of this total, 85% goes directly to the service boards according to their needs and the remainder is retained on a discretionary basis. RTA also receives from the Illinois Public Transportation Fund an amount equal to 25% of receipts realized from certain sales taxes imposed in the region by the RTA Board and collected by the Illinois Department of Revenue.

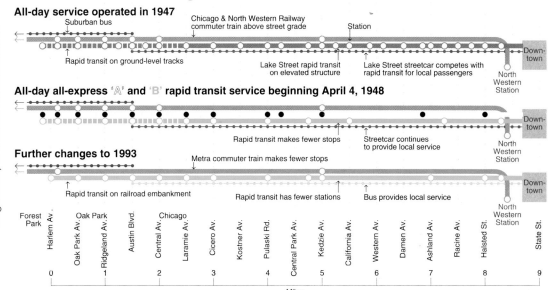

Suburban trains passing under Lake-Dan Ryan train. Metra's west line and the Lake 'L' operate through about 10 miles of a common corridor extending west to Oak Park. But Metra, which makes at most only one intermediate stop, provides weekday service of 29 train pairs (10 on Saturday, 5 on Sunday). The Lake 'L', with 9 to 11 intermediate stops, has about 190 train pairs on weekdays (well over 100 each on Saturdays and Sundays). Due to an anomaly in fare structure, at press time Metra was actually offering a lower competitive fare between Oak Park and downtown.

Rosemont/O'Hare yard - 3-18-90 - A. Peterson

The urge for heritage preservation caused then-General Operations Manager Harold H. Geissenheimer, MCERA, to have cars 6101-6102 restored to their fabulous 1950s-style mercury green, croydon cream and swamp holly orange. Out of revenue service in 1990, they remain in 1993 in work train service and are available for charter or other special occasions.

Diversey/Ravenswood line - 9-27-92 - Greg J. Sommers

When the centenary of rapid transit arrived in June 1992, CTA prepared a project, led by Bruce Nelson, of applying South Side Rapid Transit colors and graphics to Pullman 1964-built pair 2007-2008. They emerged as 1892-1992. They were used on a charter fund-raiser for the George Krambles Transit Scholarship Foundation. *(See page 68.)*

Midway/Southwest station - 6-2-91 - A. Peterson

For the Southwest Transit Project to Midway Airport the City's designers encouraged individual architectural treatment for each station while conforming to standard basic footprint and layout. The three-track terminal station at 59th Street is laid out to accommodate interchange with several bus routes, a complete yard and rail car maintenance facility. The below-grade station setup will facilitate the planned future southward extension of the rail line.

136

Integration of the transportation network of the six-county RTA area provides suburban bus by Pace to CTA railheads, as at Desplaines Transit Center. An annunciator sign on the station wall alerts bus operators to approaching trains they could not otherwise see.

On CTA, park'n'ride space was first provided in 1952 at 54th Avenue/Cermak (Cicero-Berwyn) multi-modal Transit Terminal on the Douglas branch. Beginning with the Dempster, Skokie Swift, parking lot in 1964, responsibility for operation and maintenance of lots was assumed by the local suburbs, and this has been the case at Desplaines (Forest Park) and at River Road (Rosemont).

Desplaines/Congress station - 5-5-92 - A. Peterson

1990 the combined system so recovered 55.58% of its operating cost. For the year 1990 revenue from users compared to operating costs were:

Unit	Revenues	Expenses
CTA	$388 million	$728 million
Metra	$171 million	$295 million
Pace	$ 31 million	$ 80 million

Capital expenses are funded from publicly funded grants. In 1989 the legislature authorized issuance by RTA of up to $1 billion in bonds to accelerate the extensive process of rehabilitating the region's public transit infrastructure. To date, $100 million has been issued. Prior and current investments are funded predominantly from grants provided by the Federal Transit Administration and by the Illinois Department of Transportation. To the extent of their capability, the federal and state programs have been utilized to replace buses, cars, locomotives, structures, buildings, tracks, power and distribution equipment, signals, communications, tools and a

Riverr Road/O'Hare station - 9-11-91 - A. Peterson

Kinzie/Clinton - 4-17-92 - A. Peterson

myriad of other elements. Extensions to the transit system since the 1960s have been and are being similarly financed.

As to future operations, the degree of integration in service planning, the allocation of service territory by mode, the levels of pricing and methods of fare collection between CTA, Metra, Pace and possible other players remain areas to be watched. Forty-five years of CTA background, almost twenty of it within the RTA family, have shown the need for fine-tuning from time to time and the ability to do so effectively.

The entire urbanized area of Chicago will remain in constant gradual change. Its transportation network will be ever modifying itself better to fit travel needs. To do this, CTA as the principal carrier here must continue into the 21st century to exert its leadership in transit know-how and efficiency.

◊   ◊   ◊   ◊

Taylor/Canal - 5-1-92 - M. Charnota

Over the years, cooperation between local agencies has sent a few buses, after some years of regular service, to special duties. Among the more recent were these two ex-7400 series GMC buses. At top is unit extensively modified to become the Cook County Sheriff's Emergency Services Command Vehicle, shown at work at the Kinzie Street bridge, where the terrible Chicago flood resulted from puncturing the roof of a former narrow-gage freight railway tunnel. (See also page 98.)

The lower photo shows a bus assigned to the training division of the Chicago Fire Department.

Appendix - 1947 description of CTA

The prospectus dated August 7, 1947 and issued by Blyth & Co., Inc., a member of the underwriting syndicate offering $105 million Chicago Transit Authority Revenue Bonds Series of 1947, contains a description of the properties prepared by W. C. Gilman & Company, consulting engineers. The Gilman report provides an excellent review of the transit situation in Chicago in 1947, as demonstrated by the selected portions quoted below which include a description of the properties and of the proposed modernization program.

Description of Properties

The system of the Surface Lines consists of 1,092.5 single track equivalent miles of standard gauge track with overhead trolley and 59.1 miles of double overhead trolley for trolley bus lines. Revenue passenger equipment owned consists of 3,270 street cars, 152 trolley buses and 580 motor buses. The average age of all street cars owned is 32 years, of the trolley buses 15 years and of the gas buses 4.4 years. The Surface Lines owns 15 car houses and garages…built between 1894 and 1911. In addition…there is one leased garage…

The Rapid Transit property to be acquired consists of 170.9 miles of single track equivalent of standard gauge track owned and operating rights over 70.6 miles of single track, representing a total of 87.3 miles of road. Of the total miles of single track owned 134.2 are elevated on steel structure and 36.6 miles are on the surface. The contact system includes 168 miles of third rail and 19.6 miles of overhead trolley. Passenger equipment owned by the Rapid Transit consists of 1,621 passenger cars, made up of 1,026 motor cars and 595 trailers, of which 1,473 are in regular service. The average age of these cars is approximately 42 years. Of the total, only 456 are all steel construction and 251 have steel reinforced wood superstructures and under frames. These cars are operated in trains of from one to eight cars in length.

The Rapid Transit has 235 stations… 92 of brick construction, 58 of sheet metal and wood, 31 of frame construction and the remainder of various types of construction, including 9 subway stations which are of concrete construction. The Rapid Transit has 13 car shops… Four of these are largely overhaul shops, one is entirely a body and paint shop and the remainder are inspection shops.

Both systems purchase their electric power requirements. The Surface Lines owns 20 substations and in addition receives energy from 18 substations owned by the Commonwealth Edison Company. The Rapid Transit receives power from 35 substations, only one of which is owned by the company…

The track and roadway of the Surface Lines are, in our opinion, in fair or less than average condition…there are sections on nearly all of the lines where extensive reconstruction is due now or will be due in the near future because of worn rail head, pounded rail joints or poor alignment. Over the last several years many frogs in line intersection special work have been repaired and built up by electric welding, a practice generally followed to prolong the life of special work installations. Many of these are now starting to break up and they will have to be replaced with new steel…

The car houses and garages owned by the Surface Lines, although quite old, are generally of substantial construction and have been well maintained… Facilities at… [car barns that it is planned will be retained] are being improved and modernized to permit proper inspection of PCC cars and to provide new or enlarged facilities for motor bus and trolley bus inspection and storage. Motor buses require indoor storage under Chicago climatic conditions and the facilities now generally available are the long narrow bays of existing car houses formerly containing from three to five tracks for car storage and now used for from three to five lines of motor buses. This arrangement permits practically no maneuvering of buses in storage and does not permit the greatest efficiency in the installation of facilities for fueling and oiling buses. The system needs more modern garage space for the storage of buses, particularly in view of the expected rapid increase in the future use of this type of vehicle…

While the Surface Lines shops are extensive, they do not seem readily adaptable to the efficient handling of the newer types of equipment… Uninterrupted movement between the several buildings [at West Shops] is not possible because of the intervening streets. While bus maintenance facilities are being improved and expanded at West Shops, it is felt that the expected future growth of the bus fleet will soon make these quarters inadequate. The South Shops buildings are older and no attempt has been made to modernize these facilities. Practically all the machine tools are still belt driven from overhead line shafts. Land is available at that location for expansion but the location with respect to the entire system is uneconomic unless two shop locations are to be continued.

The Rapid Transit elevated structure is generally in good condition, considering its age. Structure painting, which had to be somewhat deferred during the war, is now up to date and on schedule. While there is some structure repair work necessary at certain locations, the only major item is the reconstruction of the Wells Street viaduct over the C & N W Railway tracks on the North Side main line at Kinzie Street. This is understood to be scheduled for 1947–8. Track and track deck on the structure have been well maintained. Some of the surface track on earth elevation need reballasting and realigning.

Except for the body and paint shops at Skokie, the Rapid Transit shops are old and in very cramped quarters. The concentration of all overhaul work at one location with ample space and more modern facilities should prove more economical. Space for such a development is available at Skokie. The overall economies of the planned expansion at that site, which would involve substantial dead mileage, will have to be weighed against a new development at a more central location but probably on much more expensive land.

Rapid Transit rolling stock is largely old and outmoded. Only 28% of the passenger cars owned are of all-steel construction and all of these cars were acquired between 1912 and 1924. No passenger cars have been acquired since 1924. From a mechanical standpoint, however, the equipment has been well maintained and is in good operating condition.

The block signal system of the Rapid Transit is decidedly inadequate. Out of 208 single track miles of main line track only some 56 miles, including the State Street Subway, are protected by automatic block signals. Out of a total of 387 block signals only 139 have automatic train stops. Another 55 miles of track have spacing boards which are designed to keep a train from approaching too closely to the train ahead. The efficacy of spacing boards depends solely on the alertness of the motorman and the distance over which he has visibility. They are of little protective value in fog, smoke, snow or other atmospheric conditions which reduce visibility. Rapid Transit last year had one serious rear end collision on track protected only by spacing boards…

Modernization Program

The Transit Authority has adopted a comprehensive program for modernizing the systems to be acquired. This program contemplates expenditure of $152,374,300 and is scheduled for completion in 1955. Of this total amount $77,624,300 covers items applying to the Surface Lines system and $74,750,000 to the Rapid Transit system…

Following is a condensed, paraphrased summary of the major items announced by CTA that was included next in the Gilman report.

Item	Estimated cost (millions)
2,725 buses to replace street cars	$ 39.5
800 PCC street cars for lines retained	18.9
1,000 rapid transit cars to replace wood cars	50.0
Garages, car houses, shops for surface lines	15.0
Shops for rapid transit	1.5
Repaving, track and power for surface lines	1.2
Signal system for rapid transit	10.8
Station and structure improvement for 'L'	7.4
Repay city for Dearborn subway facilities	2.4
Structure changes to permit wider 'L' cars	1.2
Car, track and power improvement for 'L'	1.5
Total estimated cost of 10 year program	$ 152.4

From both an operating and an economic standpoint modernization of the systems is necessary. Much of the street car equipment is old and a considerable mileage of track will require reconstruction in the near future. The planned conversion by 1955 of 712 round trip miles of street railway routes (mostly two-man operation) to motor bus or trolley bus operation will permit the one-man operation of vehicles with substantial operating economies and will reduce materially the future expenditures for track reconstruction. The resulting service should be more attractive because of the more modern vehicles, and their flexibility of operation will permit better speed and a better adjustment of service to traffic requirements. It is planned that not more than 11 car lines comprising 358 miles of round trip route will continue as rail operations…

The majority of the present rapid transit cars are also old and have wooden bodies and in many cases wooden underframes. The present manpower requirements are high as an eight car train of these older cars requires eight men. All of the steel cars now owned are required for operation in the State Street Subway, where steel equipment is essential. Similarly, steel equipment will be required for the operation of the Dearborn Street Subway on its completion. The program as scheduled contemplates the replacement by 1955 of all wooden rapid transit cars with new steel equipment so designed as to expedite the handling of passengers and to permit the operation of an eight car train with three men. It also contemplates modernizing the present steel cars so that an eight car train can be operated with three men instead of five as at present. Two sample units of the new equipment are scheduled for delivery in August 1947 and each unit will be made up of three sections of articulated design mounted on 4 trucks. These sample units will have the same width at the floor as the present cars but will be several inches wider at the seat level. While we believe that complete replacement of the present wooden equipment is desirable for both passenger comfort and safety as well as for operating economies, we are of the opinion that such replacement can be carried out satisfactorily and at lower cost, either by modifications in the design of the new articulated subway cars as now proposed, by modifications in the design of present standard subway cars, or in other ways.

Included in the program is an item of $1,200,000 for structural changes which will be necessary if the final design for the new cars results in a greater width at the floor level than that of the present cars, and about $200,000 for widening of platforms of present cars to permit their operation after station platforms have been cut back. We believe that widening the cars at the seat level, as the pilot models are now designed, is sufficient and that the expenditures which would be necessitated by a widening at the floor level cannot be justified…

The Gilman report continued with analysis of revenues and costs. Basic fares in April 1947 were 9¢ CSL and 12¢ CRT, to become 10¢ and 12¢ respectively on CTA takeover. Originating riders, 1,074,878,000 in 1946, were optimistically projected to recover by regional growth after a postwar slump. Gilman concluded that the unified system "will, under good management, have available from earnings and from the assumed sale of $40,000,000 of additional revenue bonds sufficient funds to complete… programmed modernization by the end of 1957."

Bibliography

The general base of reference for the text and captions in this book is the publications of the trade press of the railway and bus transit industry plus those of the regulating, manufacturing and operating agencies for the period from the 1880s to the present, supplemented by unpublished reports, statistics and other documentation, formal and informal, relative to the subject matter.

An equally valuable resource was the collective observations and recollections of individual professionals, technicians, historians and enthusiasts.

For brevity, in the case of references drawn from a continuing series of publications, the overall source is listed, not the individual issues used.

Listings are in alphabetical order by author, except that, where no author was identified, the listing is by title.

American Car & Foundry Company
  Photographs, specifications, drawings
American Public Transit Association
  Transit Fact Book, Passenger Transport, various statistical reports
Board of Supervising Engineers of Chicago Traction
  Annual Reports, 1907-1947
Boorse, John E., Jr. - see *Cox*
Brill Company, J. G.
  Brill Magazine, catalogs, data sheets, photographs and drawings
Bus Ride (magazine)
  Friendship Publications
Campbell, George V.
  North Shore Line Memories, Domus Books, 1980
  Days of the North Shore Line, National Bus Trader, 1985
Car Builders' Dictionary
  Railroad Gazette, 1903
Carlson, Stephen P., and Schneider, Fred W., III
  PCC, the Car That Fought Back,
  PCC from Coast to Coast,
  Interurban Press, 1980, 1983
Cassier's Magazine
  The Electric Railway Number, 1899
Central Electric Railfans' Association
  Bulletins, field trips, speakers 1938 - 1992
Chicago Area Transportation Study, Final Report
  Volumes I-III, 1959
CTA Transit News
  Chicago Transit Authority, 1948 - 1993
Cincinnati Car Company
  Catalogs, correspondence, specifications, drawings
Corley, Raymond F.
  PCC cars: extensive tabulations of comparative dimensions and equipment details, 1948 - 1988

Cox, Dr. Harold E., and Boorse, John E., Jr.
  PCC Cars of North America
  Privately published, 1963
Cumella, Robert S. - see *Pushkarev*
Demoro, Harre, and Kashin, Seymour
  The PCC Car, an American Original
  Interurban Press, 1986
Dubin, Arthur Detmers
  Some Classic Trains, More Classic Trains,
  Kalmbach, 1964, 1974
Due, John F. - see *Hilton*
Electric Railroaders' Association
  Headlights, various issues, 1938-1993
  Pioneers of Electric Railroading, 1991
Electric Traction (magazine, also directory) - see *Mass Trabsportation*
Federal Transit Administration
  Statistical reports, regulations, safety studies, etc.
General Electric Company
  General Electric Review, various brochures, catalogs, data sheets, specifications, photographs and drawings, 1900 - 1990
Hansen, Zenon - see *Kristopans*
Hilton, George W., and Due, John F.
  The Electric Interurban Railway in America,
  Stanford University Press, 1960
  The Cable Car in America
  Howell-North Books, 1971
Interurbans (magazine)
  Ira L. Swett, publisher, 1957 - 1966
Johnson, James D.
  A Century of Chicago Streetcars, 1858-1958
  The Traction Orange Company, 1964
Kristopans, Andris J.; Kunz, Richard R.; McGowen, Michael M.; and Hansen, Zenon
  Chicago's Motor Coaches, Vols 1, 2 1973, 1974
Kunz, Richard R. - see *Kristopans*
Lind, Alan R.
  Chicago Surface Lines an Illustrated History
  From Horsecars to Streamliners
  Transport History Press, 1974, 1978
McGowen, Michael M. - see *Kristopans*
McGraw-Hill Publishing Company
  Electric Railway Journal (magazine)
  McGraw Electric Railway Directory
  McGraw Electric Railway Manual, 1899 - 1931
  Street Railway Journal (magazine)
  Transit Journal (magazine)
  and other publications
Mass Transportation, formerly **Electric Traction,** later **Mass Transit** (magazine, also directory and almanac), Kenfield-Davis 1910-1945, other publishers since
Metro Magazine - a Bobit Publication
Middleton, William D.
  North Shore, America's Fastest Interurban,
  Golden West Books, 1963 and numerous other books and articles 1961 - 1993

Motor Coach Age, 1958 - 1993
  Motor Bus Society, Inc.
Office of Technology Assessment of the U. S. Congress
  Automatic Train Control in Rail Rapid Transit
  also case studies of transit planning, 1976 et seq.
Plachno, Larry
  Sunset Lines / the Story of the Chicago Aurora & Elgin Railroad
  Transportation Trails, 1989
Principles of Urban Transportation
  Western Reserve University, 1951
Provenzo, Eugene F. - see *Young*
Pullman-Standard Car Company and predecessors
  Specifications, data sheets, drawings and photographs, 1892 - 1975
Pushkarev, Boris S., with J. M. Zupan and R. S. Cumella
  Urban Rail in America
  Indiana University Press, 1982
Railroad Magazine
  Various issues and dates
Railway Age (magazine) and predecessors
  Simmons-Boardman Publishing, 1880 - 1993
Richey, Albert S.
  Electric Railway Handbook, McGraw-Hill 1924
St. Louis Car Company
  Catalogs, data sheets, specifications, photographs and drawings
Schneider, Fred W., III - see *Carlson*
Sebree, (G.) Mac, and Ward, Paul
  The Trolley Coach in North America
  Interurban Press, 1974
Sprague, Frank J.
  The Multiple Unit System of Electric Railways
  American Institute of Electrical Engineers, 1899
Tuthill, John K.
  Transit Engineering
  John S. Swift & Co., 1935
Vuchic, Vukan R.
  Urban Public Transportation Systems and Technology - Prentice-Hall, 1981
Urban Mass Transportation Administration - see *Federal Transit Administration*
Ward, Paul - see *Sebree*
Weber, Harry P.
  An Outline History of Chicago Traction
  Privately published, 1936
Westinghouse Electric & Manufacturing Co., W E
  Various brochures, catalogs, specifications, data sheets and photographs, 1900 - 1990
Yerkes System of Street Railways, A History of the
  Privately published, 1897
Young, Andrew D.
  St. Louis Car Company Album
  Interurban Press, 1984
Young, Andrew D. and Provenzo, Eugene F., Jr.
  The History of the St. Louis Car Company
  Howell-North Books, 1978
Zupan, Jeffrey M. - see *Pushkarev*   ◊ ◊ ◊ ◊

Index

For your convenience in reference, six alphabetical indexes have been provided:

❶ Personal names
This index—located immediately below—lists people mentioned in the text or captions for their accomplishments.

❷ Personal credits
This index—at the right—acknowledges individuals (including photographers) who contributed to *CTA at 45*. See also the Bibliography on page 141.

❸ Locations, transit routes, stations
This index—appearing on pages 142–143—lists names of locations (including streets, transit facilities and routes) which occur in text, captions, maps or photo identification lines. Bus routes are listed by number (under "R") only if so identified in the book.

❹ Transportation providers
This index—on page 143—gives the names of transportation companies and agencies.

❺ Manufacturers, suppliers, consultants
This index—please turn to page 143—lists car and bus manufacturers and other persons and firms supplying the transit industry.

❻ Other keywords
This index—also on page 144—contains keywords which do not fall into any of the above categories.

Also see the table of contents on page 3.

---

❶ Personal names
Listed here are editorial references to people (*for acknowledgments, please refer to index ❷*).

Adams, John Quincy, *10*
Anderson, Bruce, *93*
Arnold, Bion J., *19*, 111
Aurand, J., 10
Banks, Ernie, 10, *11*
Barnes, Eugene M., 10, *11*
Belcaster, Robert P., 5, 10
Brabec, Edward F., 10
Brady, Michael I., 10
Budd, Britton I., 107
Budd, Ralph, *8*, 10, 31
Burrus, Clark, 10
Cafferty, Michael J., 10, 99
Cardilli, Michael A., 10
Charlton, James I., 10
Clark, Walter, 10
Collins, Philip W., *10*, *11*
Daley, Richard J., *11*
Daley, Richard M., 123
Delgado, Natalia, 10
DeMent, George L., 10, *33*
Donenfeld, J. Douglas, 10
Dreiser, Theodore, 16
Dubin, Arthur D., *128*
Egan, J. E., *11*
Elkins, 16
Fallon, Bernard J., 111
Ferguson, Edgar, *85*
Ford, Bernard J., 10
Fox, Kim B., 10
Fronczak, Steven, *105*
Gallagher, James P., 10, *11*
Gasior, Andrew J., *105*
Geissenheimer, Harold H., *136*
Getz, George F., Jr., *10*
Gunlock, Virgil E., 10, *11*
Harrington, Philip, *10*
Hill, Arthur F., Jr., 10
Hill, T., 10
Hillman, Jordan J., 10
Hoellen, John J., 10
Holmes, C. B., 13, 17
Holmes, John, 10, *11*
Holzman, Milton, 10
Hughes, Tommy, *85*
Humiston, John F., *58*, *80*
Hunter, Ronney, Jr., *105*
Insull, Samuel, 109, 111

Jakubowski, Mathilda A., 10, *11*
Janousek, O. F., *11*
Janssen, William C., 111 n.
Johnson, J. F., 13
Johnson, Wallace D., 10
Keevil, Charles E., 65
Klug, Charles, *117*
Kole, P., 10
Krambles, George, 10, *11*
Mason, Roswell B., 12
McCarter, Walter J., *5*, 10, 85
McDonough, James J., 10, *11*
McKenna, William W., *8*, *10*, *11*
McMillan, Howard, 77
McNair, Frank, 10
Medley, Howard C., Sr., 10, *11*
Miller, John S., 10
Moore, Edward F., 10
Murphy, Joseph D., 10
Nelson, Bruce, *136*
Newell, Lem, *101*
O'Connor, Thomas B., 10
Paaswell, Robert E., 10
Parmelee, Frank, 12–13
Pavoni, Walter, *85*
Peacock, Raymond D., 10
Petzold, Charlie, *122*
Phillips, Charles B., 12
Phinger, Jennifer, *117*
Pikarsky, Milton, 10
Porter, Irving L., *8*, *10*
Powell, Fred, *101*
Quinn, James R., *10*, *11*
Reyes, Guadalupe, 10
Richardson, Guy A., 10
Ritchie, John A., 44
Roddewig, Clair M., 10
Ruggiero, Nick, 10, *11*
Rutherford, James E., 10
Savage, Alfred H., 10
Schroeder, Werner W., 10
Schuster, Theodore G., 10
Sprague, Frank Julian, 24 n., 59, 63
Sucsy, Lawrence C., 10
Tracy, Ralph W., 86
Traiser, Louis M., 85
Van Der Vries, Bernice T., 10
Walsh, Donald, 10
Ward, Jim, *52*
Widener, Peter A. B., 16
Winston, Ollie, *85*
Yerkes, Charles T., 12–13, *16*–18, 20, 24

---

❷ Personal credits
Listed here are acknowledgments of people. The photographers' credits also appear alongside each photo. (*See also the Bibliography, page 141.*)

Andersen, Glenn M., *34*, *131*
Bartkowicz, Ronald, 4
Benedict, Roy G., 4, *6*, *61*, *78*, *102*, *103*, *119*, *130*, *135*
Benton, Robert L., 4
Bisset, Kendrick D. G., 4
Borchert, Fred, *14*, *106*
Buckley, James J., *15*, *57*
Campbell, George V., *28*
Carlson, Norman, 4
Charnota, Mike, *54*, *116*, *138*
Culbertson, Lillian, 5
Darling, John A., 4
Desnoyers, Thomas H., *16*, *20*, *30*, *31*, *32*, *108*
Diaz, Joe L., *14*
Dubin, Arthur D., 4
Frank, Ed, Jr., *106*
Hirsch, Harold R., 5
Janssen, William C., 4, *13*. *32*, *42*, *91*
Kadowaki, Paul, 4
Kehl, Roy A., 5
King, Fred G., 4
Krambles, George, 3, 4, *8*, *10*–12, *14*–16, *18*, *19*, *21*, *23*–*31*, *33*–*36*, *40*, *42*, *44*–*48*, *55*–*57*, *60*, *62*, *70*, *71*, *75*, *76*, *78*, *84*–*86*, *88*, *89*, *91*, *92*, *95*–*97*, *107*, *109*, *110*, *114*, *125*, *128*, *132*, *133*, *dust jacket*
Lloyd, Arthur L., 4
Mehlenbeck, Robert V., *17*, *38*, *90*
Offett, Celestine, 5
Peterson, Arthur H., 3, 4, *37*, *49*–*54*, *58*, *64*, *66*, *69*, *72*, *74*, *77*, *79*–*83*, *87*, *98*–*101*, *104*, *105*, *112*, *113*, *117*, *122*, *127*, *134*–*138*, *dust jacket*
Peterson, Sarah, dust jacket
Peterson, Sean, dust jacket
Peterson, Stephen, dust jacket
Peterson, Tina, dust jacket
Randich, Gene M., 4
Santoro, John, 5
Savage, Alfred H., 5
Smerk, George M., 4
Sommers, Greg J., *136*
Stone, Barney L., *34*
Walter, Jim, 4
Wolgemuth, Thomas L., 4

---

❸ Locations, transit routes, stations
(Except as noted, streets are in Chicago)
Many street and municipality names, not individually indexed, appear in the map on page 6.

5th Av., *39*
5th Av. (now Wells St.), *106*, 108
11th St., *18*
12th St. (now Roosevelt Rd.), 12, 13
14th Blvd., *18*
18th St., *18*, 72, 74, *81*, 123
21st St., 16
22nd St. (now Cermak Rd.), 12
22nd St. (Westchester, IL), *34*, 108
23rd St., *18*
31st St., *18*
35th St., 96, *120*
37th St., *18*
39th St. (now Pershing Rd.), 21, *22*
47th St., *20*, 43
49th St., 123
51st St., 5, 43
52nd St., *70*
54th Av. (Cicero, IL), *78*, *137*
55th St., 5, 43
56th St., *54*
59th St., 72, 123, *136*
61st St., 72, *78*, 82, 94, *99*
63rd Pl., *38*, *107*
63rd St., *23*, *25*, *47*, *70*, 106, *107*, 109
69th St., *32*, *49*, *54*, *78*, 96, 121
77th St., *51*, *54*, *78*, *133*
78th St., *75*, 77
79th St., *21*, *53*, 76, 130
87th St., *19*
88th St., 107
95th St., 20, *51*, 65, 94, 96, *120*, 121
98th St., *78*
103rd St., *33*, *51*, *78*
111th St., 20
Adams St., *25*, 96
Addison St., *16*, 122
Albany Av., *114*
Archer Av., 12, *19*, *46*, *51*, *78*, 107, 123
Armitage Av., *17*, 43, *61*
Ashland Av., *135*
Aurora (IL), 107 n., 108
Austin Blvd., *26*, *96*, *135*
Barry Av., 71
Batavia (IL), 108
Bellwood (IL), *106*, 108
Belmont Av., *42*, 43, 122
Berwyn Av., *29*
Beverly garage, *33*, *75*, *78*
Bishop St., *51*
Bissell St., *64*
Blue Island (IL), 106, 107 n.
Blue Island Av., 17
Boston (MA), 59, *60*, 66, 66 n.
Bosworth Av., see Cooper St.
Brandon Av., *15*
Britain, 50
Broadway, *82*
Brooklyn (NY), *60*, 61; see also New York (NY)
Buena Av., 111
Buffalo (NY), *104*
Burnham Park, *20*
California Av., 43, *135*
Calumet Av., *70*, 72
Calvary Cemetery, 21
Canada, 100
Canal St., *138*
Carroll Av., *80*
Central Area Circulator, see index ❻
Central Av., *39*, *42*, 43, *46*, *135*
Central St. (Evanston, IL), *24*, 29
Central Park Av., *40*, *48*, *135*
Cermak Rd., 12, 20, *27*, *30*, 121; see also 22nd St.
Cermak Rd. (Cicero, IL), *137*
Charlotte (NC), *51*
Chicago, *throughout book*
Chicago Av., 24, 43
Chicago Heights (IL), 107 n.
Chicago River, *14*, 16, 17, 73, 98, 118, 124
Cicero Av., *21*, *43*, *78*, 96, 107, 108, *135*
Cicero-Berwyn Terminal, *137*
Cincinnati (OH), 56
Clark St., *90*, *101*
Cleveland (OH), *60*
Clinton St., *17*, 96, *135*, *138*
Clybourn Av., *12*, 27, *81*
Columbus Dr., 74
Congress route, 93, 96, *98*, 113, 120, 122, *137*; see also West-Northwest route

Congress (now Eisenhower) Expy., 71 n., 73, *97*, 118
Congress St./Pkwy., 21, 27, *31*, *33*, *43*, *60*, 71, 73, *97*, *108*, 116, 118
Cook County (IL), 7, 30, 38, 118, 118 n., 128, 134 n., 135 n.
Cooper St. (now Bosworth Av.), *12*
Cottage Grove Av., 12, *13*, *23*, *30*
Crawford Av. (now Pulaski Rd.), 45
Cumberland Av., 123
Damen Av., *135*
Dan Ryan route, *37*, 38, *54*, 65, 72, 73, 74, *81*, 94, 95, 96, *120*, 121, 121 n., *122*, 123, *135*; see also North-South route, West-South route
Davis St. (Evanston, IL), *93*, 94
Dearborn St., 12, 27, *33*, 61, 61 n., 73, *97*, *98*, 116, 118, 120, 140
Dempster St. (Evanston, IL), 94
Dempster St. (Niles Center/Skokie, IL), *29*, 66, 73 n., 109, 125, 126, *127*, *128*, *137*
Desplaines Av. (Forest Park, IL), *78*, 98, 108, 112, 113, 118, *137*
Devon Av., *45*, 46
Diversey Av., 20, *42*, 43, 45, 122, *136*
Division St., 27, *98*
Dorchester Av., 74, 109
Douglas route, 71, 71 n., 73 n., 93, *98*, 119, 120, 122, *127*, *137*; see also West-Northwest route
Düsseldorf (Germany), *50*   [route
DuPage County (IL), 38, 134 n.
East Chicago (IN), 20, 106 n.
East Prairie Rd. (Skokie, IL), 126
East St. Louis (IL), 38
Edison Building, *84*
Eisenhower Expy., 118; see also Congress Expy.
El Cajon (CA), *104*
Elgin (IL), 107 n., 108
Elm St., 20
Elmwood Park (IL), 7
Elston Av., *16*, 43, 76
Englewood route, 24, 27, *32*, *37*, *47*, 72, *107*, 121; see also North-South route
Europe, 67
Evanston (IL), 8, 21, *24*, 27, 29, 29 n., *37*, *55*, *58*, *60*, 61, 63, 67, 72, 73 n., 74, 92, *93*, 94, 109, 110, 113, 115, 125
Evergreen Av., 73
Ewing Av., *15*
Federal St., 74, 123
Florence Av. (Evanston, IL), *69*
Forest Glen garage, *52*, *75*, *78*
Forest Park (IL), 118–119, 120, *135*
Fort Sheridan (IL), 111
Foster Av., *46*, *50*, 122
Franklin St., 12, 20
Fullerton Av., 12, 18, 43
Garfield Blvd., *55*
Garfield Park route, 5, 108, 118, 119 n., 120
Gary (IN), 123 n.
Geneva (IL), 108
Germany, 50, *51*, 67, 126
Grand Av., *35*, 43, *97*
Granville Av., 115
Great Lakes Naval Training Station (IL), *110*
Greenleaf Av. (Wilmette, IL), *110*   [111
Grove St. (Evanston, IL), *58*
Guadalajara (Mexico), *53*
Halsted St., 18, *27*, *47*, *49*, 106, *107*, 118, *135*
Hammond (IN), 20, 106 n.
Harlem Av., *52*, *122*, 123
Harlem Av. (Forest Park/Oak Park, IL), 73, *78*, Harrison St., *49*, 71 n., 79   [94, 96, *135*
Harvard Av., *32*
Harvey (IL), 106
Hawthorne Training Center, *78*, *99*, *105*
Hegewisch, 15, 123
Hermitage Av., *50*, *51*, *52*, *88*, *113*
Higgins Av., *122*
Hobbie St., *14*
Holden Ct., *31*, *79*

Howard St., 21, 27, *29*, *37*, 38, *51*, *52*, *66*, *71*, 74, *78*, *82*, 82 n., *88*, 94, 109, *113*, 121, 125; see also North-South route
Hubbard St., *18*
Illinois, 7, 37, 38, 65, 104, 118, 127, 135 n., 137
India, 100
Indiana, 38
Indiana Av., 72
Interstate 55, 120, 123
Interstate 90, 120, *122*, 123
Interstate 94, 120, 123
Interstate 190, 123
Interstate 290, *66*, 118
Interstate 294, 123
Irving Park Rd., *28*, *29*, 43, *76*, 122
Jackson Blvd., *97*
Jackson Park, 21, *25*, 27, *37*, 72, 74, *99*, 121; see also North-South route
Jefferson Park station, 38, *61*, 65, 121 n., 122,
Joliet (IL), 107 n.   [122 n., 123, *134*
Kane County (IL), 38, 134 n.
Kankakee (IL), 106, *107*
Kedzie Av., *40*, 43, 49, *52*, *54*, *78*, *135*
Kenmore Av., *91*
Kennedy Expy., 65, 73, *120*, 121, 121 n., 122
Kenosha (WI), 109
Kenton Av., 98, *115*, 118–119
Kenwood route, 24
Kimball Av., 21, 43, *78*, *83*, 94, 122
King Dr., see South Park Av.
Kinzie St., 98, *138*, 139
Kostner Av., *135*
Kostner Av. (Skokie, IL), 126
Lake St., 13, *22*, 24, *26*, *27*, 29, 30, 32, *34*, *37*, 55, *57*, 71, 71 n., *72*, 73, *78*, *79*, 82, 83, *85*, 87, 92, 93, 95, *96*, *109*, *110*, *114*, *115*, *116*, 119, 120, 121, *135*; see also West-South route
Lake Bluff (IL), 110–111
Lake County (IL), 38, 128, 134 n.
Lake Michigan, 109
Lake Park Av., *54*
Lake Shore Dr., *20*
Laramie Av., 45, *78*, 94, 108, 118, 119 n., *135*
Larrabee St., *32*
LaSalle St., 16, 17, 17 n., *33*, 71, *97*
Latrobe Av., *26*
Lawndale garage, *78*, 104
Lawrence Av., 43
Leavitt St., 123
Lehigh Av., *46*
Lexington St., *46*
Libertyville (IL), 110
Limits garage, *78*, *101*
Lincoln Av., 24
Linden Av. (Wilmette, IL), *58*, 74, 82, 94
Llewellyn Park (now Wilmette, IL), 29 n.
Lockwood Av., 119
Logan Square, *33*, *62*, 71, 71 n., 73, *97*, 120,
Lombard (IL), 112   [121, 122
Long Av., *34*
Loomis St., 119
Los Angeles (CA), dust jacket
Loudonville (OH), *49*
Madison St., 12, 17, *22*, *39*, *40*
Main St. (Skokie, IL), 126, *127*
Manhattan (NY), 44, 61; see also New York (NY)
Mannheim Rd., *66*
Mannheim Rd. (Westchester, IL), *34*, 108
Market St. (now Wacker Dr.), *22*, 26
Marshfield Av., *108*
Maypole Av., 83
Maywood (IL), 5, 108
McHenry County (IL), 38, 134 n.
Merchandise Mart, *35*, *80*, *85*, *86*, *88*
Metropolitan West Side Elevated Railroad/ Railway Co., *23*, 24, 57, *106*, *108*, *110*
Michiana Regional Airport (IN), 123 n.
Michigan Av., *48*, 124
Midway Airport, 38, *55*, 67, *74*, 107, 123, *136*
Milwaukee (WI), *109*, 110, 121
Milwaukee Av., 17, *18*, 27, *33*, *46*, 73, *98*, 115–116, 118, 120, 122

Mohawk St., *77*
Montana St., *79*
Montrose Av., 29, 43, *76*, 82, *91*, 111, 122, 128
Morgan St., 119
Mt. Carmel Cemetery, *106*, 108
Mundelein (IL), 110–111
Narragansett Av., 43
Navy Pier, 124
Nelson St., *80*
New York (NY), 28, 44, *46*, 59; see also Brooklyn (NY), Manhattan (NY)
Niles Center (now Skokie, IL), 27, *29*, 109, *125*; see also Skokie (IL)
Normal Park route, 24, *32*
Norristown (PA), *53*, 60
North Av., 43, *77*, *89*, *90*, *131*
North Park garage, 49, *51*, *54*, *75*, *78*
North Water St., *11*, *31*
North Western Station, *46*, 123, *135*
North-South route, 8, *32*, 73, 74, *81*, *82*, 87, 93, 94, 96, *99*; see also Dan Ryan route, Englewood route, Howard St., Jackson Park
North-South Tollway, 123
Northwest Hy., *50*
O'Hare International Airport, 38, *64*, *66*, 73, *98*, 120, 121, 122, 122 n., 123, 123 n., *124*, 128, *134*, *136*, dust jacket; see also West-Northwest route
O'Harexpress route 40, *50*, 122 n.
Oak Park (IL), *26*, *34*, 119, 120, *135*
Oak Park Av., *38*
Oak Park Av. (Oak Park, IL), *135*
Oak Ridge Cemetery, *106*
Oakton St. (Niles Center/Skokie, IL), *2*, 79, *125*,
Oriole Av., *2*   [*126*, *dust jacket*
Park Av., see South Park Av.
Park Ridge (IL), *112*
Parkside Av., *30*, *34*
Parnell Av., *32*
Paulina St., 71, 72
Pershing Rd., see 39th St.
Philadelphia (PA), 16, 17, *38*, 59, *60*, 65
Portugal, 67
Prairie Av., *55*, *70*
Prairie Rd., see East Prairie Rd.
Pratt Av., *52*
Pulaski (WI), 109   [Crawford Av.
Pulaski Rd., *21*, 43, 96, 122, *135*; see also
Racine (WI), 109
Racine Av., *60*, *78*, 119, *135*
Randolph St., 12, 13, *36*, *48*, 116, 123
Ravenswood route, 27, *35*, *58*, *60*, *61*, 61 n., 72, 73 n., *80*, *83*, 93, 94, *127*, *136*
Ravenswood Av., *44*
Red Lion (PA), 65, 67
Richmond (VA), 17, 24 n.
Ridge Av., *46*
Ridgeland Av. (Oak Park, IL), *135*
River Rd. (Rosemont, IL), 73, 122, 123, 123 n.,
Rockwell St., *19*, *46*, 71, *78*, *114*   [*137*
Rogers Av., *50*
Roosevelt Rd., *18*, 20, 43, 74, 109, 123; see also 12th St.
Rosemont (IL), *64*, *78*, 123, 123 n., *136*, *137*
Routes by bus/streetcar route number appearing in text; see also street/route names:
 —Route 4 Cottage Grove, *30*
 —Route 8 Halsted, *47*
 —Route 16 Lake, 55, *115*
 —Route 17 Westchester, 113
 —Route 20 Madison, *39*
 —Route 40 O'Harexpress, *50*, 122 n.
 —Route 51 Sheridan, 45
 —Route 61 Archer/Franklin, 107
 —Route 62 Archer, 107
 —Route 99 Stevenson Express, 107
 —Route 157 Streeterville, 51
 —Route 303 Madison/19th, 113
 —Route 310 Madison/Hillside, 113
Sacramento Av., *57*
Sacramento Blvd., *57*
St. Charles (IL), 108

St. Louis (MO), 28, 38
St. Mary's of the Lake (IL), 110
San Diego (CA), *104*
San Francisco (CA), 13, 17, 66, 66 n.
São Paulo (Brazil), 69
Schubert St., *101*
Sedgwick Av., *58*
Sheffield Av., 71, *79*, *80*
Sheridan Rd., *28*, 44, *45*, *46*
Skokie (IL), *58*, 79, 80, 81, 125, 126, 139; see also Niles Center (IL)
Skokie Shops, *78*, 79, 80, *95*, 125, *127*
Skokie Swift, *2*, 21, *37*, *55*, *60*, 61 n., *63*, *66*, *67*, *69*, 73 n., 74, 88 n., 92, 93, 96, 113, *125–128*, *137*, dust jacket
Soldier Field, *100*, *133*
South Bend (IN), 123 n.
South Park Av. (now King Dr.), 106, 107
South Shops, *53*, 76, *78*, 80, 133, 139
Southwest route, see Midway Airport
Spain, 50
State St., 12, 13, *18*, *27*, *31*, *33*, *36*, 44, *56*, 59, *60*, 61, 61 n., 73, *81*, *87*, 96, *98*, 115, *135*, 140
Stevenson Expy., 120, 123
Stewart Av., *32*
Stock Yards route, 24, *27*
Stony Island Av., *19*, *25*, *33*
Streeterville route 157, 51, 124
Summit Av. (Park Ridge, IL), *112*
Sweden, 50, 100
Taylor St., *138*
Throop St., *60*
Tonty Av., *46*
Touhy Av. (Park Ridge, IL), *112*
Tower 12, 72, *74*
Tower 18, *72*, *74*, *109*
Troy (NY), 13
Union Av., *107*
Union Station, 44, 123
University Av., 94
Van Buren St., 16, 17, 17 n., 72, *74*
Vincennes Av., *33*, *75*, 76, *77*, 106, 107, *133*
Wabash Av., *22*, *25*, 45, 72, *74*, *79*, *87*, 109, *116*, 123
Wacker Dr., *16*, *22*; see also Market St.
Washington Blvd., *48*
Washington St., *17*, 17 n., *22*, *31*, *33*
Water St., see North Water St.
Waukegan (IL), 29 n., 109, 112
Wells St., *31*, *72*, *109*, 139; see also 5th Av.
Wentworth Av., *75*, *81*
West Shops, *78*, 83, 139
West-Northwest route, 73, 128; see also Congress route, Douglas route, O'Hare International Airport
West-South route, 72, 94, 96; see also Dan Ryan route, Lake St.
Westchester (IL), 5, 27, *34*, 108
Western Av., 20, *21*, *42*, *53*, 94, *135*
Wheaton (IL), 112
Will County (IL), 38, 134 n.
Willow St., *46*
Wilmette (IL), 27, 29 n., *58*, 67, 74, *110*, 111
Wilson Av., *23*, 24, *29*, *37*, *44*, *57*, 74, *78*, *82*,
Wolf Lake, *15*   [93, 94, 111
Zaire, *53*

❹ Transportation providers

Aurora Elgin & Chicago Railway, *106*, 108, 109
Baltimore & Ohio Chicago Terminal Railroad, 118
Bay Area Rapid Transit District, 66
Bi-State Development Agency, 38
Bluebird Coach, 29
Calumet & South Chicago Railway Co., 7, *15*, 39, 130 n.
Calumet Electric Street Railway, *15*
Central Area Circulator, see index ❻
Chicago & Calumet District Transit Co., 20
Chicago & Evanston Railroad, 21
Chicago & Interurban Traction Co., *107*
Chicago & Joliet Electric Railway, 107
Chicago & Milwaukee Electric Railroad/Railway, 29 n., 109
Chicago & North Western Railway, *26*, *34*, 80, 83, 110, 121 n., *135*, 139
Chicago & South Side Rapid Transit Railroad Co., 21
Chicago & West Towns Railways, 29, *132*
Chicago & Western Railway Co., 7
Chicago Aurora & Elgin Railroad/Railway, *108*, 111, 112–113, 119 n.
Chicago Burlington & Quincy Railroad, *31–32*
Chicago City Railway Co., 7, *13*, 16, 17, 20, *38*, 39, 106–107, 130 n.
Chicago Electric Traction Co., 106
Chicago Elevated Railways Collateral Trust, 26, 130 n.
Chicago Fire Department, *138*
Chicago Junction Railway, *27*
Chicago Milwaukee & St. Paul Railway, *24*, 29 n., 110
Chicago Milwaukee St. Paul & Pacific Railroad (Milwaukee Road), 8, *29*, 94 n., 111
Chicago Motor Bus Co., *28*, 44
Chicago Motor Coach Co., 7–8, *17*, *18*, 28, *44–45*, *46*, 46 n., *48*, *132*
Chicago North Shore & Milwaukee Railroad/ Railway (North Shore Line), *27*, *29*, 60, 109–113, *109*, *110*, 111 n., 115, 125, *133*
Chicago Railways Co., 7, 20, 39, 130 n.
Chicago Rapid Transit Co., dust jacket
 —cars of, 56, 59, 61
 —description of, 12, 139–140
 —development of, 21–27
 —engineering on, 104, 111, 111 n.
 —intercompany transfers, *132*
 —power distribution of, *77*, *84*
 —purchased by CTA, 7–8, *10*, 28
 —relations with Chicago Motor Coach Co., 28
 —service supervision of, 85
 —signal system of, *30*, 115
Chicago South Shore & South Bend Railroad (South Shore Line), 38, 111, 111 n., 123
Chicago Stage Co., 44
Chicago Surface Lines:
 —bus service of, 28
 —cars and trolley buses of, 20, *38*, 39, 39 n., *40*–43
 —debt of, 9
 —description of, 12, 45, 139–140
 —development of, 13–21
 —effect of subway construction on, *18*
 —engineering on, 61, 104
 —motor buses of, 45
 —night service of, 28
 —power distribution of, *77*
 —purchased by CTA, 7–8, 28
 —relations with other transportation providers, 44, *107*
 —service supervision of, 84–85
 —snow removal by, *90*
 —unusual traffic on, 111 n.

Chicago Transit Authority, throughout book
 —acquisition of North Side rapid transit by, 29
 —Anthon (H. S.) Memorial Library of, 5
 —board of, 7–8, *10*, *11*
 —chief executives of, 5, 7, 10, dust jacket
 —engineering on, 9
 —formation of, 1, 7–8, dust jacket
Chicago Tunnel Co., 98
Chicago Union Traction, *114*
Chicago West Division Railway, 13
Cook County, 107
Cook County & Southern, 106
Cook County Sheriff, *138*
Depot Motor Bus Lines, 44
Evanston Bus Co., 113, *133*
Fifth Avenue Coach, 44, *46*
Greyhound, *134*
Hammond Whiting & East Chicago Railway, 107
Illinois Central Railroad, *13*, *18*, 107
Illinois Railway Museum, *21*, *42*
Indiana Railroad, dust jacket
Lake Shore & Michigan Southern Railway, 13
Lake Street Elevated Railroad Co., *22*, 24
Leyden Motor Coach, 29
Metra, 38, 123, 124, 129, *134*, *135*, 138
Metropolitan West Side Elevated Railroad/ Railway Co., *23*, 24, 57, *106*, 108, *110*
Milwaukee Road, see Chicago Milwaukee St. Paul & Pacific Railroad
Minneapolis St. Paul & Sault Ste. Marie Railway, 110
North Chicago Street Railroad/Railway Co., *12*, 13, 16, 17
North Shore Line, see Chicago North Shore & Milwaukee
North Suburban Mass Transit District, 113
Northern Indiana Commuter Transportation District, 38, 111 n., 123, 123 n.
Northwestern Elevated Railroad Co., *23*, 24, 29, 29 n., *82*
Nortran, 113
Omnibus Corp., The, 28
Pace, 29, 38, *113*, 129, *134*, 135, *137*, 138
Regional Transportation Authority (RTA), 8, 29, 38, *112*, 129, 134–135, *137*, 138, dust jacket
San Diego Trolley, *104*
South Shore Line, see Chicago South Shore & South Bend Railroad
South Side Rapid Transit, *13*, *22*, *23*
South Suburban Safeway, *133*
Southeastern Pennsylvania Transportation Authority, 53
Southern California Rapid Transit District, dust jacket
Southern Street Railway, 7, 39, 130 n.
Union Consolidated Elevated Railway Co., 7
Union Elevated Railroad Co., 24
United Motor Coach Co., *113*
Washington Metropolitan Area Transit Authority, 66

More indexes appear on the next page.

143

❺ Manufacturers, suppliers, consultants

Alsthom, ***64***
AM General, ***51***
American Public Transit Assn., 98, ***100,*** 129, dust jacket
Baldwin Locomotive Works, ***29,*** 57
Blyth & Co., Inc., 139
Boeing-Vertol, 65–67, ***99***
Brill (J. G.) Co., ***20, 26, 38,*** 39, ***131***
Budd Co., ***64,*** 65, 67
Central Electric Railfans' Assn., 5, dust jacket
Clark Equipment Co., ***60,*** 61, 63
Commonwealth Edison Co., ***14,*** 139
Cummins, ***54***
Detroit Diesel, ***54***
Diamond-T, ***77***
Dodge, ***21***
Dubin, Dubin & Moutoussamy, ***128***
Duewag, 69
Eaton, Gilbert & Co., 13
Electric Railway Presidents' Conference Committee, 40
Flxible Corp., ***49, 52, 54,*** 55
Ford, ***21***
General Electric, 61, 65, 67
General Motors Corp., ***48, 52, 53, 115, 138***
Gilman (W. C.) & Co., 139–140
GM Truck & Coach, 49, 55
IBM, ***132***
ICF Kaiser Engineers, dust jacket
International Union of Public Transport, dust jacket
Jewett, ***23, 26***
Krambles (George) Transit Scholarship Fund, 4, dust jacket
Krupp, 67
LFM Co., 65
Mafersa, 69
MAN, ***50, 51, 52***
Marmon-Herrington, ***43***
Matra Transit, 123, dust jacket
Morrison-Knudsen Co., 67
Motorola, 87 n.
Muller-Munk, Peter, 65
New York Rail Car Corp., ***64,*** 67
PKS Kiewit Western, ***127***
Public Service Company of Northern Illinois, 109
Pullman, ***60,*** 64, 67, 116
Pullman Car & Manufacturing Co., 64–65
Pullman-Standard, 61
Rhode Island Locomotive Works, 57
Rockwell Co., 65
Rohr, 66
St. Louis Car Co., ***26, 28, 42,*** 44, ***60,*** 61, 62
Skidmore Owings & Merrill, 65, ***120***
Sorefame, 67
Sundberg-Ferar, 66
Transit America, 67
Transportation Manufacturing Corp., ***52, 54,*** 55, ***115***
Transportation Research Board, dust jacket
Twin Coach, ***42, 43,*** 45, ***49***
University of Chicago, 100
Walsworth Publishing, 4
Walter (Jim) Color Separations, 4
Wegmann, 67, 69
Westinghouse, ***29,*** 61, 63
White Motor Co., ***21,*** 45, ***46***
Yellow Coach, ***45***

Page references given in plain type refer to the text.
Page references shown in ***bold italics*** refer to illustrations or their captions or to photo credits.

❻ Other keywords

'A' and 'B' skip-stop express service, ***32, 57,***
Accidents, *see* Collisions 63, ***72,*** 84, 103
Air conditioning, ***52,*** 55, ***61,*** 65, 69, 120, 127
Alley 'L,' 5, 21
Amalgamated Transit Union, 105
Americans with Disabilities Act of 1990, 53
Articulated vehicles, *see* Motor buses, Rapid transit cars, Trolley buses
Automatic block signals (ABS), *see* Signals
Automatic Vehicle Monitoring (AVM), 88–89
Bicentennial, *see* Color schemes
Block signals, *see* Signals
"Bluebirds," 60
Board of Supervising Engineers, Chicago Traction, ***19***
Bonds, revenue, 7–***8,*** 8 n., 28, 29, 31, 112, 139
Boulevard Route, the, 28
Brakes, *see* Rapid transit cars—brakes, Trolley
Bus shelters, 77, 83 buses—brakes
Buses, *see* Motor buses, Omnibuses, Trolley
Cab signals, *see* Sig buses
Cable cars, *see* Streetcars, cable
Car washing, ***31,*** 82, 82 n.
Carhouses, ***16, 75,*** 139, 140
Cars, *see* Hospital cars, PCC cars, Rapid transit cars, Smoking cars, Streetcars
Central Area Circulator, 80, 123–124, dust jacket
Century of Progress exposition, ***18, 27***
Changes in potential traffic, 1, 5, 9, 33, ***35,*** 36–37, 119 n., 121, 129, dust jacket; *see also* Competition by autos
Charter service, ***106, 133, 136;*** *see also* Inspection trips
Chicago Area Transportation Study, 125, 128
Chicago Department of Public Works, ***120, 122***
Chicago Department of Subways and Superhighways, 61, 61 n.
Chicago Department of Transportation, ***122***
Chicago Transit Authority Technical Institute,
Coal, ***14, 29,*** 111, 99–100, 103
Collisions, 49, 80, 90, 114, 115, ***116,*** 117, 140
Color schemes, ***47, 48, 49, 55, 112***
—heritage, ***58, 80, 136***
Commuter ("suburban") trains, ***6,*** 7, 13, 38, ***135;*** *see also* Metra *in index* ❹
Competition:
—by autos, 1, 9, ***21,*** 35, 36, 62, 84, 129, 132,
—by other transit providers, 7, 13, ***15, 17,*** 28,
Crossings, *see* Grade crossings
Curvature in track, ***22,*** 59, ***79***
Defense plants, ***21***
Depots, *see* Carhouses
Derrick car, ***23***
Destination signs, 23, ***52,*** 55, 63–64, 65
Dreamland dance hall, ***108***
Electric buses, *see* Trolley buses
Electric locomotives, ***29,*** 111
Electric power facilities, 12, ***14,*** 34, 40, ***77, 84,*** 85, ***86, 88,*** 137, 139; *see also* Underground
Electric streetcars, *see* Streetcars conduits
Electroliners, 60, ***109***
Elevated structures, ***22, 25, 27,*** 75, ***78, 79, 80,*** 82, 83, 91, ***114, 115,*** 139; *see also* 'L,' Rapid transit, Stations
Employee Assistance Program, 100–101
Epidemics, ***23***
Equal employment opportunity, 100
Equipment trusts, 32
Equity in service delivery, 100
Express service, *see* 'A' and 'B'
Fairs, *see* Century of Progress exposition, World's Columbian Exposition
Fares, 1, 9, 13, 26, 30, 36, 40, 103, 130–133, ***131,*** 132 n., ***135,*** 140
Federal Transit Administration (FTA), ***66,*** 67, 125
Fires, ***27,*** 94
Floods, 90, ***97, 98, 138***
Fluorescent lighting, ***47,*** 121
Fog, ***91,*** 140
Freight service, ***29,*** 107, 111–112, 111 n.
Funeral trains, ***106***
Future of transit, 5, 138
Garages, ***16, 19, 33, 75,*** 76, 78, 88 n., ***89,*** 91, 139, 140; *see also names of specific garages in index* ❸
Gauntlets, ***24, 29***
Generating stations, *see* Electric power facilities
Grade crossings, 73–74, 73 n., 80, 94 n., ***105,*** 117
Grants, 8, 137
Graphs:
—accidents per 100,000 vehicle-miles, ***117***
—CTA fares, ***130***
—trend of transit ridership nationwide, ***129***
Heritage (historic) equipment, ***58, 80, 136***
Horse cars, *see* Streetcars, horse
Hospital cars, 107, 110
Housing and Home Finance Agency, 125, 126
Human relations, 100
Ice, 63, 90, 92, 93, 94, 95
Identra, 71 n., 72 n.
Illinois Commerce Commission, ***132***
Illinois Department of Transportation, 37, 67
Illinois Institute of Technology, dust jacket
Inflation, 9, 35, 36, ***130,*** 132
Influenza, ***23***
Inspection trips, 5; *see also* Charter service
Interlocking, *see* Signals
International Bus Roadeo, ***100***
International Eucharistic Congress, 109–111, 111 n.
Interurban railways, 29 n., ***106–110,*** 115, 125, 126; *see also* Electroliners
Kiss 'n' ride, 126
'L' (defined), 9 n.; *see also* Elevated structures, Rapid transit, Stations
Left-hand operation, 26
Licenses, car, ***26***
Light rail transit, ***104,*** 123, 128; *see also* Central Area Circulator
Lighting, fluorescent, ***47,*** 121
Lill Coal Co., ***29***
Locomotives:
—electric, ***29,*** 111
—steam, 5, 21, ***22, 23,*** 24, 57, 59
Loudspeakers, *see* Public address systems
Management development, 100, 104
Management Institute, 100
Maps: west Chicago, ***119***
—improved rapid transit service for north-
—local and through transit markets in one corridor, ***135***
—principal maintenance facilities, ***78***
—transit service at CTA's inception, ***6***
—where employees live, ***102***
—where employees report to work, ***103***
Marian Year observance, ***133*** 39 n.
Metropolitan Transit Authority Act, 7, ***10,*** 28,
Metropolitan Water Reclamation District, 31
Minority/disadvantaged business enterprise, 100
Motor buses; *see also names of manufacturers in index* ❺; *see also* Color schemes *in this index*
—articulated, ***50–51***
—compressed natural gas (CNG), 49
—conversion to, ***19,*** 20, 21, ***30,*** 31, ***34,*** 45, ***46,*** 75, 125, 140
—diesel, 45, ***46,*** 47, 49, 55, 103
—double-deck, ***28, 44–45, 46,*** 50
—gasoline, 40, 44, 45, ***46,*** 47, 47 n., 139
—introduction of, 28, 44, 45
—lift-equipped, ***52, 54,*** 55
—"New Look," 48, 53
—propane, ***36,*** 47–***49,*** 103
—size of, ***48, 49,*** 50, 55
—transmissions, 45
—work buses, ***49***
National Housing Act of 1961, 125
Northeastern Illinois Planning Commission, 125
Omnibuses, 12–13
One-person operation, conversion to, 9, ***19, 30,*** 34, ***44, 55, 69,*** 140
Open cars, *see* Streetcars, summer
Operating stations, *see* Carhouses, Garages
Operation Control Center, 84, ***85, 86,*** 87, ***88,*** 89
Ordinances, municipal, 7, 18, ***19,*** 20, 26, 28
Overhead wire, *see* Trolley wire
Painting, 31, 79–80; *see also* Color schemes
Park 'n' ride, 123, 126, 128, ***137***
PCC cars, 21, ***30, 38, 39, 40–41, 60,*** 61, 62, 103, 139, 140
People mover, 123,
Police, 24, 26, 76, 110, 114, ***117***
Power plants, *see* Electric power facilities
Private-sector transit, 1, 7, 9, 20, 134, dust jacket
Public address systems, ***25, 44,*** 55
Public Utility Holding Company Act, 129 n.
Public Works Administration, 27
"Queen Mary" buses, 44, ***45***
Rails, 121, 139
Rapid transit, operation by steam locomotives, 5, 21, ***22, 23,*** 24, 57, 59
Rapid transit cars; *see also* Color schemes, Heritage equipment, Locomotives, PCC cars:
—1–50 series, 63, 65, 67, 79 n.
—60 series (cars 61–65), 67, 69
—2000 family (cars 2001–3456), ***66;*** *see also* 2000 series, 2200 series, 2400 series, 2600 series, 3200 series *below*
—2000 series (cars 2001–2180), ***61, 64–***65, 69, ***96***
—2200 series (cars 2201–2352), ***64,*** 65, 66, 67, 69, ***96***
—2400 series (cars 2401–2600), 65–67, 67 n., 69
—2600 series (cars 2601–3200), 67, 67 n., 69, 79
—3200 series (cars 3201–3456), 67, 67 n., ***68–69,*** 123
—4000 series, ***56, 57, 58,*** 59, 63, 63 n., 65,
—5000 series, ***60,*** 61, 61 n., 67 92
—6000 series/family, ***61, 62–***63, ***64,*** 65, ***66,*** 67, 79 n., ***80***
—all-steel, ***55, 56, 57,*** 59, 59 n., ***110,*** 139,
—articulated, 56, ***60,*** 61, 126 140
—"baldies," 59; *see also* 4000 series *above*
—brakes, 57, ***58,*** 59, 94
—derrick car, ***23***
—descriptions of, 63–67
—development of, 57–69
—door control in, ***55,*** 59, ***62,*** 63
—electric locomotive cars, 21, ***22,*** 24, 57
—high-performance, ***11,*** 63, 64, 126
—hospital, 110
—introduction of multiple-unit control, 21, ***23,*** 24 n., 59, 59 n.
—married pairs, ***60,*** 61, ***62,*** 126
—motor, 56, ***57, 63,*** 139
—"plushies," 59; *see also* 4000 series *above*
—State of the Art, ***66***
—trail, ***23, 56, 57,*** 63, 139
—trolley equipment on, 59, 63, 70, ***110,*** 126,
—utility cars, ***80, 136, 127***
—wooden, 12, 56–***57, 58,*** 59, 63, ***110,*** 115, 139, 140
Rapid transit stations, *see* Stations
Revenue bonds, 7–***8,*** 8 n., 28, 29, 31, 112, 139
Seat arrangements, 55, 56, 56 n., ***57,*** 59, 65 n.
Securities & Exchange Commission, 129 n.
"Sedan" streetcars, 40
Semaphore signals, *see* Signals
Shelters, bus, 77, 83
Shops, 12, 81–82, ***83,*** 121, 123, 139–140; *see also names of specific shops in index* ❸
Signals, ***17, 23,*** 26, ***30,*** 32, ***58,*** 59, 65, 65 n., ***69, 70–72,*** 71 n., 73, 73 n., ***74, 80, 91,*** 103, ***108,*** 115–116, 126, 137, 140
Signs, 23, ***52,*** 55, 63–64, 65, 77
Skip-stop service, *see* 'A' and 'B'
Sleet, 90, 92
Smoking cars, ***23***
Snow removal, ***90–***98, ***92, 93, 96***
Spacing boards, *see* Signals
State of the Art cars, ***66***
Stations, operating, *see* Carhouses, Garages
Stations, rapid transit, 12, ***22,*** 119, ***120,*** 121, 122–123, ***136,*** 139
Storerooms, 83
Street crossings, *see* Grade crossings
Street paving, ***19,*** 34, 140
Streetcars; *see also* PCC cars:
—cable, ***12, 13, 14,*** 16–17, 20
—compressed air, 17
—descriptions of, 39–40
—discontinuance of, 62; *see also* Motor buses, conversion to
—gas, 17
—horse, 13, 16, 20
—hospital, 107
—introduction of electric, ***15,*** 17, 18, 20
—last extension of, ***20***
—mail, ***14***
—open, ***15***
—"Sedans," 40
—soda, 17
—steam, 17
—summer, ***15***
—utility cars, 39
Structures, *see* Bus shelters, Carhouses, Electric power facilities, Elevated structures, Garages, Shops, Storerooms, Terminals
Substance abuse, *see* Employee Assistance Program
Substations, *see* Electric power facilities
Suburban trains, *see* Metra *(in index* ❹*),* Commuter trains *in this index*
Subway construction, ***18,*** 27, ***33,*** 122; *see also* Rapid transit, Stations
Supertwin bus, ***42,*** 43
Surface system, *see* Motor buses, Streetcars, Trolley buses
Taxes, 8, 36, 67, 135, 135 n.
Terminals, off-street, 77, 91, 121
Third rail, ***23, 24, 58, 70, 87,*** 92, 93, 94, 139
Through routing of rapid transit, ***23,*** 26, 38
Ties, 121, 121 n.
Track maintenance/renewal, 34, ***81,*** 82, 83, 121 n., ***127,*** 139, 140
Tracks, gauntlet, ***24, 29***
Training, ***99–105***
Transmissions, *see* Motor buses—transmissions
Trolley buses, 21, ***39,*** 40, ***42–43, 46,*** 47, ***53, 76,*** 139
—articulated, ***42***
—brakes, 43
—conversion to, 21, 40, ***42,*** 43, 103, 140
—discontinuance of, 21, 43
Trolley equipment, *see* Rapid transit cars—trolley equipment on
Trolley wire, 18, ***26,*** 45, ***58, 70, 76,*** 92, 139
Trucks:
—car, 56, 65, 67, 69
—road, *see* Utility vehicles
Tunnels (for streetcars), ***16, 17,*** 17 n.
Underground conduits, 18, ***77***
Undeveloped territory, ***15, 21, 34,*** 123, ***125***
Union Stock Yards, ***27***
Unions, labor, ***30,*** 81, 105
U. S. Dept. of Commerce, 92 125
U. S. Dept. of Housing & Urban Development,
U. S. Dept. of the Interior, 27
U. S. Dept. of Transportation, 120, 121, 125
University of Illinois, dust jacket
Urban Mass Transportation Administration, 125
Utility (work) vehicles, ***23,*** 39, ***49,*** 76, ***77, 80,*** 82, 84, ***86, 136***
Vans, *see* Utility vehicles
Vehicle Maintenance System, 76
Washing cars, ***31,*** 82, 82 n.
Weather Bureau, 91, 92
Wire, *see* Trolley wire
World's Columbian Exposition, ***13, 25***
World's fairs, *see* Century of Progress exposition, World's Columbian Exposition
Yards, 12, 118, 118 n., 122; *see also names of specific terminals in index* ❸

◊ ◊ ◊ ◊

603